Simulating Society

Richard J. Gaylord   Louis J. D'Andria

Simulating Society

A *Mathematica*® Toolkit for Modeling
Socioeconomic Behavior

Richard J. Gaylord
Department of Materials Science and Engineering
University of Illinois at Urbana-Champaign
Urbana, IL 61801
USA

Louis J. D'Andria
Wolfram Research, Inc.
100 Trade Center Drive
Champaign, IL 61820-7237
USA

Library of Congress Cataloging-in-Publication Data
Gaylord, Richard J.
    Simulating society : a Mathematica toolkit for modeling
  socioeconomic behavior / Richard J. Gaylord, Louis J. D'Andria.
        p.   cm.
    Includes bibliographical references and index.
    ISBN 0-387-98532-8 (pbk. : alk. paper)
      1. Social interaction—Computer simulation.   2. Socialization-
  -Computer simulation.   3. Mathematica (Computer file)   I. D'Andria,
  Louis J.   II. Title.
  HM291.G375   1998
  302'.0285—dc21   98-17543

Printed on acid-free paper.

Production managed by Steven Pisano; manufacturing supervised by Jacqui Ashri.
Photocomposed pages prepared from the authors' *Mathematica* notebook files.
Printed and bound by Hamilton Printing Co., Rensselaer, NY.
Printed in the United States of America.

9 8 7 6 5 4 3 2 1

ISBN 0-387-98532-8 Springer-Verlag New York Berlin Heidelberg SPIN 10659681

THE
ELECTRONIC
® LIBRARY
OF
SCIENCE

TELOS, The Electronic Library of Science, is an imprint of Springer-Verlag New York. Its publishing program encompasses the natural and physical sciences, computer science, mathematics, economics, and engineering. All TELOS publications have a computational orientation to them, as TELOS' primary publishing strategy is to wed the traditional print medium with the emerging new electronic media in order to provide the reader with a truly interactive multimedia information environment. To achieve this, every TELOS publication delivered on paper has an associated electronic component. This can take the form of book/diskette combinations, book/CD-ROM packages, books delivered via networks, electronic journals, newsletters, plus a multitude of other exciting possibilities. Since TELOS is not committed to any one technology, any delivery medium can be considered. We also do not foresee the imminent demise of the paper book, or journal, as we know them. Instead we believe paper and electronic media can coexist side-by-side, since both offer valuable means by which to convey information to consumers.

The range of TELOS publications extends from research level reference works to textbook materials for the higher education audience, practical handbooks for working professionals, and broadly accessible science, computer science, and high technology general interest publications. Many TELOS publications are interdisciplinary in nature, and most are targeted for the individual buyer, which dictates that TELOS publications be affordably priced.

Of the numerous definitions of the Greek word "telos," the one most representative of our publishing philosophy is "to turn," or "turning point." We perceive the establishment of the TELOS publishing program to be a significant step forward towards attaining a new plateau of high quality information packaging and dissemination in the interactive learning environment of the future. TELOS welcomes you to join us in the exploration and development of this exciting frontier as a reader and user, an author, editor, consultant, strategic partner, or in whatever other capacity one might imagine.

TELOS, The Electronic Library of Science
Springer-Verlag New York, Inc.

THE
ELECTRONIC
® LIBRARY
OF
SCIENCE

TELOS Diskettes

Unless otherwise designated, computer diskettes packaged with TELOS publications are 3.5″ high-density DOS-formatted diskettes. They may be read by any IBM-compatible computer running DOS or Windows. They may also be read by computers running NEXTSTEP, by most UNIX machines, and by Macintosh computers using a file exchange utility.

In those cases where the diskettes require the availability of specific software programs in order to run them, or to take full advantage of their capabilities, then the specific requirements regarding these software packages will be indicated.

TELOS CD-ROM Discs

For buyers of TELOS publications containing CD-ROM discs, or in those cases where the product is a stand-alone CD-ROM, it is always indicated on which specific platform, or platforms, the disc is designed to run. For example, Macintosh only; Windows only; cross-platform, and so forth.

TELOSpub.com (Online)

Interact with TELOS online via the Internet by setting your World-Wide-Web browser to the URL: *http://www.telospub.com*.

The TELOS Web site features new product information and updates, an online catalog and ordering, samples from our publications, information about TELOS, data-files related to and enhancements of our products, and a broad selection of other unique features. Presented in hypertext format with rich graphics, it's your best way to discover what's new at TELOS.

TELOS also maintains these additional Internet resources:

gopher://gopher.telospub.com
ftp://ftp.telospub.com

For up-to-date information regarding TELOS online services, send the one-line e-mail message:

send info

to: *info@TELOSpub.com*.

To my life companion, Carole. The one thing I could never throw away is my feeling for you. And to a kindred spirit, Calvin of Calvin and Hobbes.

—R. J. G.

To Suzanne, for the inspiration, support, and love she cheerfully and incessantly gave during and in spite of the creation of this book.

—L. J. D.

Although agent-based modeling employs simulation, it does not aim to provide an accurate representation of a particular empirical application. Instead, the goal of agent-based modeling is to enrich our understanding of fundamental processes that may appear in a variety of applications. This requires adhering to the KISS principle, which stands for the army slogan "keep it simple, stupid."

—Robert Axelrod

Foreword

Richard Gaylord and Lou D'Andria's book is not a standard introduction to computational social sciences filled with long passages on the history of economic thought. Instead, this is a book describing how to create computer models of complex human interactions. These models stand in sharp contrast to the static, "supply equals demand" models of introductory economics, though that is not to say that standard economic models are not informative, just that they are predictable. In contrast, the models presented in this book can exhibit perpetually novel behavior.

A computational model relies on individual agents (people) who have preferences and can take actions. Agents can walk. They can move to new neighborhoods if they feel uncomfortable in their present situation. They can play games with their friends, or choose to hang with a new crowd. They can share gossip. They can chat on the phone. While these agents are reasonably smart, they do not always optimize. To borrow Herb Simon's words, they are boundedly rational. They go about life much in the same way that we do. They rely on heuristics—rules of thumb—in their day to day interactions. These rules are not set in stone. In fact, if an agent finds that she is using a stupid rule, she may abandon it in favor of something better.

In watching and analyzing the microlevel and aggregate behavior of these artificial agents, we can learn much about our own world. We learn that gossip has its good side and that the presence of racial segregation need not imply that individual agents are racist.

This latter finding is due to Thomas Schelling, whose book *Micromotives and Macrobehavior* still informs those of us interested in dynamic models. Gaylord and D'Andria perform a much needed service by making this model available to all. They also extend Schelling's model by allowing agents to relocate based on endogenous (acquired) characteristics. For example, I may choose to move to another house if my neighbor decides to form a rock band, or I may choose to move out of town if my fellow citizens vote to cut public services or raise taxes to a level I find objectionable.

The Gaylord and D'Andria book carves out an important niche in this growing field of complexity, owing to its accessibility. Although many models of artificial society make their computer codes available publicly, these often consist of thousands of lines of inelegant code. By relying both on *Mathematica*, a high-level language, and on exacting standards, the authors construct elaborate models with minimal, often beautiful, code. For example, some of their models require less than fifty lines of code. Gaylord and D'Andria also deconstruct their code and discuss how to extend it to accommodate richer models. Finally, their code produces readable, analyzable output. They reveal that with a few simple subroutines, some courage, moderate intelligence, and a little heart, anyone can create artificial societies.

This book contains models that are both bottom-up and complex: the accumulated behaviors of autonomous agents generate outcomes, and agents respond to an environment that they in turn define by their actions. The more advanced models in the book include adaptive behavior: agents abandon unsuccessful behavioral rules. Herein lies the difference between social science models and those from the physical sciences—people can change. We can copy strategies of others, alter our own actions, or act out of caprice. Carbon atoms never consciously decide to throw caution to the wind and act like helium atoms. Yet poor people sometimes act rich, smart people sometimes act foolish, and conservative people sometimes take huge risks. Although this book only dips its toe into the water on the issue of how to construct artificial adaptive agents, it does provide a sufficient introduction to get people started.

The topics addressed in this book—movements, fads, norms, game playing, social networks, culture, and conformity—span traditional social scientific boundaries. Many of the models take as a fundamental assumption that others—friends, family, and peers—influence our actions. Recently, several economists, notably Steven Durlauf of the Santa Fe Institute and the University of Wisconsin, have begun to analyze the role of social influence in explaining persistent social problems such as crime, drug use, out-of-wedlock births, and school attendance. These problems are difficult to analyze empirically, so simulations and other forms of theorizing seem a logical first step in organizing our collective thinking.

This book was not written with the intent of influencing social policy directly and, given its intent, it wisely avoids long discussions on patterns of drug use, crime, and other "heavy" topics. Instead, it provides an introduction. It says, "Here is how to do make your own models." The task of interpretation is left to the readers. We can attach weight to the insights we gather—asking, for example, whether these models say anything about why crack became so popular so fast—or we can just have fun.

In sum, this book merits a careful read. For the noneconomist, this book should whet your appetite for a deeper look at social science models. For social scientists who use *Mathematica* for the sole purpose of solving algebraic problems, here's a chance to see its awesome power.

Gaylord and D'Andria have included many exercises and extensions, some of which are open research questions. I encourage readers to go beyond these and formulate their own models, to think more broadly. Explore. Take risks. Construct models of product competition, movie demand, marriage markets, political preferences, religious affiliation, tax evading, investment strategies, traffic jams, or joke telling. Given the many movie quotes contained throughout the text, I close by paraphrasing a certain Mr. Gump. This book is "like a box of chocolates." You really never do know what you might get. That's the wonder of computational modeling. You define the world, then sit back and watch it evolve.

Scott E. Page
Department of Economics
University of Iowa

What This Book Is About

Our purpose in writing *Simulating Society: A Mathematica Toolkit for Modeling Socioeconomic Behavior* is two-fold:

- To demonstrate the ability of computers to create simulations of the behavior of people in social and socioeconomic situations, to people interested in human social behavior, including professionals working in the fields of social science, economics, and political science; graduate and undergraduate students (and even high school students) studying these subjects; and amateur science enthusiasts for whom the study of human society is an avocation; and

- To provide people with computer-based tools that will allow them to carry out their own computer simulation studies of whatever socioeconomic phenomena interest them.

Computer simulation studies are widely used in the physical sciences as an adjunct to theoretical studies. In the social sciences where the inherent complexity of human socioeconomic behavior limits the use of traditional mathematical analysis, simulation studies can provide an alternative theoretical tool for exploring specific aspects of human behavior.

In this book, we focus on cultural exchanges and socioeconomic transactions that occur as a result of interactions between individuals, rather than on behavior imposed by a central authority. This approach is called *bottom-up* or *agent-based* modeling.

Several general features distinguish agent-based models from traditional models of socioeconomic behavior.

- The characteristics of agents can be heterogeneous, each one having its own characteristics (identities, traits, preferences, tastes, memories).

- The reasoning of agents can be rational or irrational, intelligent or naive.

- The interactions between agents can be direct.

- The behaviors of agents can change over time.

- The location of agents can shift as they move around the system.

In addition to these general features, the models discussed in this book have these specific characteristics:

- the movement of agents can occur simultaneously or asynchronously;

- the interaction of agents can be unilateral or multilateral;

- the behaviors of agents can change within a single generation as they learn from experience and adapt accordingly.

Note: Mechanisms of change occurring over a number of generations of agents involving the use of processes that eliminate and/or create agents by allowing them to die and to give birth or genetic algorithms to mimic the evolution of the agent population, would not be difficult to incorporate into our models.

One of the major features of this book is that the code in our simulation programs is developed and explained in detail in the text. This is done for two reasons.

- We feel that the main purpose of scientific research is the creation of public knowledge, hence, published experimental, theoretical, and simulation results must be presented in such a way as to permit other researchers to attempt to duplicate the results. In the case of simulation studies, we are especially concerned with the GIGO (garbage in, garbage out) effect, particularly when long, often difficult to trace, computer programs are involved. It is therefore essential in our view, that a complete description of a program, as well as the program itself, be presented and discussed along with any results it produces.

- This book is intended, as the subtitle indicates, to be a practical guide that will enable the reader to carry out his or her own socioeconomic simulation studies, rather than to deal with so-called *deep* issues related to the principles of socioeconomic modeling. Therefore, the details of the programs need to be explicitly explained so that the programs can be understood and if desired, modified by the users for their own work.

Note: The results of the particular model simulations runs that are presented here are NOT meant to be used to evaluate the usefulness of the models themselves; that would require an in-depth study of the robustness of the results, such as their sensitivity to the program parameter values, which we do not undertake here.

Intimately related to the issue of presenting and explaining code is the choice of programming language. Because our book is meant to serve as a toolkit that readers can use to explore their own ideas of social behavior by computer simulation, we needed to choose a programming language that is powerful in its capabilities, yet simple in its use, and able to be run on a large number of computing platforms.

We have found that for writing, running, and analyzing computer simulations, the integrated computing environment provided by *Mathematica* has a number of features that make it uniquely qualified for this task, including the following.

- The *Mathematica* programming language is very simple in principle, employing a pattern-matching, term-rewriting style that allows programs to be written directly and easily. Moreover, *Mathematica* is an interpreted language, and therefore the code that makes up a program can be created and debugged in a piecemeal fashion, greatly decreasing the time needed to create a new program. Just as important, an existing program can be modified and extended, both easily and rapidly, to incorporate more complications. *Mathematica* is also a very easy language to learn, requiring no prior programming experience.

- *Mathematica* can manipulate data structures (e.g., lists and matrices) in their entirety (e.g., shifting the positions of all of the elements of a matrix or performing arithmetic operations on all of the corresponding elements of several matrices at once). In addition, *Mathematica* has extensive built-in functions for performing list manipulations such as adding, dropping, and replacing list elements, rearranging the elements, and extracting elements from the list, which allows us to work with groups of nested list structures. These allow us to write programs that are both quite concise and very readable.

Note: To enable readers who are unfamiliar with *Mathematica* to understand the code in the book, and also to write their own simulation programs, there are appendices explaining how *Mathematica*'s programming language works, and demonstrating its list manipulation capabilities. In fact, we **STRONGLY** recommend that the reader, regardless of prior experience with *Mathematica*, read through the appendices before turning to the chapters in the book, as much of the code used in the chapters is explained and illustrated in the appendices.

- *Mathematica* has extensive graphics which are quite easy to use. As a result, many different types of graphical representation can be made to find the most informative graphical format to use for a particular simulation model.

Note: Some of the features that have been introduced in version 3.0 of *Mathematica* allow one to create and display an animation while the program is running which is particularly useful for simulation work.

- *Mathematica* has extensive numeric capabilities that allow statistical analysis to be used to assign various quantitative measures (i.e., numerical values) to the simulation results.

Note: Although we do not carry out any substantial statistical analysis here, focusing instead on the creation of the simulation programs that can be run to generate data, there are statistical packages available in *Mathematica* that can be used to analyze these data.

- *Mathematica* programs are executable on a very great number of computing platforms, essentially any platform that runs *Mathematica*.

All of these features are available in an integrated computing environment, in which simulation programs can be written, debugged, run, visualized, analyzed, and even written up for publication or presentation in a single document, known as a notebook, which is very convenient. In fact, this book has been written entirely as *Mathematica* notebooks. Readers interested in viewing these notebooks or using them as a starting point from which to develop their own simulation programs should visit the TELOS web site, http://www.telospub.com.

Based on all of these factors, we have found that *Mathematica* is a nearly ideal computing environment for carrying out our investigations.

Each chapter in this book is organized as follows.

- We begin with an introductory discussion of the socioeconomic behavior(s) being modeled. Since we do not claim to be professional social scientists (our expertise resides in simulation modeling techniques and in *Mathematica* programming), we keep our discussion of background rather brief. A short list of useful, readable references to both the popular and technical literature is given at the end of the chapter. Although the references are limited to those sources that were directly relevant to the development of the models in this book, the references that these sources contain can, in turn, be used by the interested reader to delve more deeply into the subject.

- We develop programs for simulating these behaviors in a step-by-step manner with each step explained. This exposition constitutes the main body of the chapter. By first describing a model and then implementing it in a computer program while providing a detailed discussion of the program code, we prepare the reader for the final part of the chapter, where he or she can participate.

- We suggest a number of simulation projects that modify and/or extend the models in the chapter. These projects can be launchpads for readers to start their own explorations of whatever ideas they might have on human socioeconomic behavior.

References

Our selection of the social and socioeconomic phenomena that are modeled in this book, which include *social learning* via cultural or meme transmission, and *social capital* via various versions of the prisoner's dilemma, neighborhood formation, conformism, and social networking, has been intentionally eclectic so that each reader can find at least one, and perhaps several, topics of personal interest.

Our own starting point for learning about a number of these topics came from reading a popularized science book (and then the references given therein) which provides an excellent introduction for the layperson,

> Ridley, Matt. 1997. *The Origins of Virtue: Human Instincts and the Evolution of Cooperation*, Bergenfield, NJ: Viking.

An interesting collection of technical academic studies of some of the same sorts of socioeconomic phenomena that we discuss here can be found in

> Arthur, Brian, Steven Durlauf and David Lane. 1997. *The Economy as an Evolving Complex System II*. SFI Studies in the Sciences of Complexity, vol. XXVII, Reading, MA: Addison-Wesley.

Two other texts using the agent-based approach to socioeconomic modeling have been published while this book was in preparation:

> Axelrod, Robert. 1997. *The Complexity of Cooperation: Agent-Based Models of Competition and Cooperation*. Princeton, NJ: Princeton University Press.
> Epstein, Joshua M. and Robert Axtell. 1996. *Growing Artifical Societies: Social Science from the Bottom Up*. Cambridge, MA: Brookings Institute Press/MIT Press.

Acknowledgements

We would like to express our appreciation to Allan M. Wylde who was our publisher at TELOS from the inception of this project until it entered the production stage. His recognition of the potential importance of this book and his unflagging support and enthusiasm for it were essential. RJG would like to thank Allan personally for working on his behalf over the course of the years on this and three previous books. The work of the always-friendly and ever-efficient Keisha Sherbecoe at TELOS is also much appreciated.

Jerry Lyons and Steven Pisano at Springer-Verlag became directly involved with the authors during the production process, and by facilitating the creation of the final book, helped bring it into the world. Other helpful people at Springer-Verlag include Karen Phillips, who worked with us to create a really fine-looking front cover, and Fred Bartlett, who helped with the production of the *Mathematica* graphics.

The reviewers made a number of very good comments and suggestions that improved the presentation in the book and we thank them for that. We would also like to thank the numerous people from many different fields including physics, economics, political science, social science, film criticism, and cardiology who expressed their belief in the value of this project.

We would especially like to thank Scott Page for providing encouragement to TELOS to support this book and for suggesting several of the computer simulation projects at the end of various chapters. We also greatly appreciate his writing a foreword for the book.

Richard J. Gaylord and Louis J. D'Andria

Contents

Appendices

Introduction

"I'm walking here! I'm walking here!"

—Dustin Hoffman in *Midnight Cowboy* (1969)

1

Modeling a Society of Mobile Heterogeneous Individuals

"We don't go anywhere. Going somewhere is for squares. We just go."

—Marlon Brando in *The Wild One* (1954)

1.1 Introduction

This chapter describes a model of individuals moving about on a lattice. The system is *decentralized* in that there is no central authority deciding which individuals will remain where they are, which individuals will move, and to where they will move. Instead, on each time step, every individual on the lattice attempts to move at the same time (this is known as synchronous updating) and the movement is local in the sense that an individual can only move from his current lattice site location to an adjacent site and not to a site farther away. We want to avoid having individuals bumping into one another in the process of moving and since there is no traffic controller to coordinate the movements of people, two requirements are imposed on each individual that must be met in order for him to make a move, as opposed to remaining in place: the adjacent site that the individual is attempting to move to must not already be occupied by another individual and the move must not cause a collision with another individual trying to move onto the same site. As we show, we are able to express the rules governing movement subject to these conditions in a set of 28 rules, each having a very simple form. In subsequent chapters (except in the final chapter where nonlocal movement to faraway sites is allowed), these rules are used to incorporate the movement of individuals into other models of socioeconomic behavior since the mobility of individuals is a central feature in many human activities.

1.2 Milling About

A system consisting of interacting individuals, each with his own characteristics (e.g., cultural values, race, resources), can be simulated using a *discrete dynamical system* where space, time, and the states of the system are all discrete and have the following properties.

- Space is represented by a regular two-dimensional lattice.

- Each site, or *cell*, in the lattice is in a given state at a given instant of time. The states of lattice sites can be integers, reals, symbols, or strings, as well as tuplets (lists) with elements of any type.

- The system *evolves* over a succession of *time steps*. The values of all of the sites in the lattice are *updated* synchronously in each time step.

Note: Although the values of all of the sites are updated in each time step, the value of a site need not change in a given time step.

- A site's value is updated using a set of rules (known as a *lookup table*) which takes the values of the site and other sites into account.

Note: Several rule sets (which need not take the values of the same set of sites into account) may be applied successively, one after the other, to the lattice during a given time step.

1.2.1 The System

Our model uses a square n by n lattice with wraparound boundary conditions (so that some of the neighbors of a site on a border of the lattice are taken from the opposite border of the lattice). There is a population density p of individuals occupying lattice sites and the remaining sites stay empty. The system evolves over a given number of time steps t.

1.2.2 Populating Society

We first need to specify the values of the lattice sites:

- An empty site has a value of zero.

- A site occupied by an individual has a value which is a list. The elements of the list represent the characteristics of the individual. These values may change during a time step.

Note: In writing the update rules for a lattice-based model of mobile individuals, it is useful to focus on the lattice sites rather than on the individuals. For example, rather than say that an individual moves, it is better to say that a site that is occupied becomes empty and a site that is empty becomes occupied by an individual.

We can give an example of a configuration of a system of individuals by creating an n by n lattice containing a density p of individuals. Each individual is described by a list containing these quantities:

- the direction the individual is facing, indicated by 1, 2, 3, or 4, representing north-facing, east-facing, south-facing, and west-facing, respectively. (This is discussed in more detail shortly);

Note: This is done by randomly generating an integer between 1 and 4.

- a unique name (tag) identifying the individual, represented by a number (1, 2, 3, ...);

Note: This is done by setting the initial value of the symbol k to 0, and then incrementing k by 1 prior to using it to assign a name to an individual.

- the behavior used by the individual when interacting with other individuals (0 representing bad behavior and 1 representing good behavior);

Note: This is done by randomly generating a 0 or 1.

- a set of culturally determined beliefs, values, or attitudes, represented by a list of s values, where each value is an integer ranging from 1 to m.

Note: This is done by creating a list of s elements, each being a randomly generated integer between 1 and m.

- a resource level, represented by an integer value between 0 and r.

Note: This is done by randomly generating an integer between 0 and r.

The code for creating this particular system is given by (this code is developed and fully explained in Appendix B).

```
k = 0;
RND := Random[Integer, {1, 4}];
society = Table[Floor[p + Random[]], {n}, {n}] /.
  1 :> {RND, ++k, Random[Integer],
      Table[Random[Integer, {1, m}], {s}],
      Random[Integer, {0, r}]}
```

Take as an example, a 4 by 4 lattice, having 25% of its sites occupied by individuals, of whom 75% are good guys, having a set of 3 attributes ranging in value from 1 to 10 and resource levels between 0 and 4.

```
n = 4; p = 0.25; g = 0.75; m = 10; s = 3; r = 4;
k = 0;
```

```
SeedRandom[8]
RND := Random[Integer, {1, 4}];
society = Table[Floor[p + Random[]], {n}, {n}] /.
  1 :> {RND, ++k, Floor[g + Random[]],
        Table[Random[Integer, {1, m}], {s}],
        Random[Integer, {0, r}]} // MatrixForm
```

$$\begin{pmatrix} \{4,\ 1,\ 0,\ \{10,\ 8,\ 5\},\ 0\} & 0 & 0 & 0 \\ 0 & \{2,\ 2,\ 0,\ \{9,\ 4,\ 4\},\ 1\} & 0 & \{4,\ 3,\ 1,\ \{8,\ 2,\ 8\},\ 4\} \\ 0 & 0 & 0 & \{3,\ 4,\ 1,\ \{2,\ 2,\ 1\},\ 1\} \\ 0 & 0 & 0 & 0 \end{pmatrix}$$

1.2.3 Executing a Time Step

The time step is executed in two or more consecutive *partial-steps*, one involving the movement of the individuals and the others involving interactions between individuals. In each partial-step, a set of rules is applied to each site in the lattice. The arguments of the rules are the value of a site and the values of other selected sites.

The sites whose values are used in the rules are taken from amongst the sites that are in the vicinity, or *neighborhood*, of one another, such as the following.

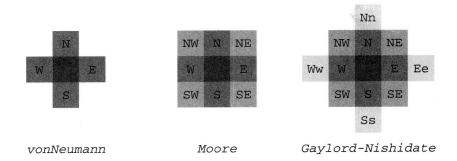

vonNeumann   Moore   Gaylord-Nishidate

Note: In the models that follow in all but the final chapter of the book, the only sites that are considered in writing the rules for updating the value of a given site during a time step are in the neighborhood of that site (in the final chapter, we use sites as rule arguments based on their either being occupied by certain individuals or being empty regardless of their location on the lattice). This type of system is known as a cellular automaton, of which the Game of Life cellular automaton is probably the best-known example.

Moving

In a system of individuals moving about in space, the most fundamental restriction that must be placed on their movement is that no more than one individual can be on a given lattice site at a given time. This is known as the *excluded volume constraint*.

On each time step, we allow each individual on the lattice to move to the nearest neighbor site he is facing and choose a random direction to face, with the following exceptions:.

- If the nearest neighbor site is occupied by another individual, the individual remains in place and chooses a random direction to face.

- If the site is empty but is faced by one or more other individuals on its nearest neighbor sites, the individual remains in place and chooses a random direction to face.

As we stated earlier, it is best to focus on the lattice sites rather than on the individuals. Recasting the preceding description in terms of the sites, we can say:

on each time step, the following events happen.

- Each lattice site that is occupied by an individual becomes empty unless (a) the nearest neighbor site faced by the individual is occupied by another individual, in which case the site remains occupied by the individual who chooses a random direction to face; or (b) the nearest neighbor site faced by the individual is empty but is faced by one or more other individuals, in which case the site remains occupied by the individual who chooses a random direction to face.

The exceptions covered by (a) and (b) are illustrated in the following.

North-Facing Walker Facing an Empty Site Faced by One or More Other Walkers

Walker Facing Another Walker

- Each lattice site that is empty becomes occupied by an individual who chooses a random direction to face unless (c) the empty site is faced by two or more individuals on adjacent sites, in which case it remains empty; or (d) the empty site is not faced by any individuals, in which case it remains empty.

The exceptions covered by (c) are illustrated in the following:

Empty Site Faced by Two or More Walkers

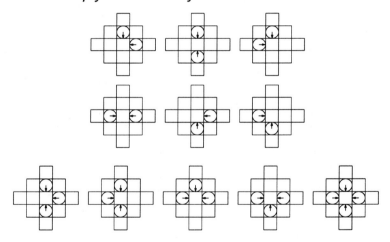

As can be seen from the preceding diagrams, updating the lattice sites under this avoidance strategy depends on the values of 12 neighboring sites (called the Gaylord–Nishidate or GN neighborhood). Accordingly, the walk rules for updating a lattice site have the form

```
walk[site, N, E, S, W, NE, SE, SW, NW, Nn, Ee, Ss, Ww]
```

where the 13 arguments of the walk rule represent the value of the site, the values of the four nearest neighbors in the N, E, S, W directions, the values of the four nearest neighbors in the NE, SE, SW, NW directions, and the values of the four next nearest neighbors in the N, E, S, W directions.

The site values used in the moving rules represent the following quantities.

0 — an empty site

{1, a___} — a site occupied by a north-facing walker

{2, a___} — a site occupied by a east-facing walker

{3, a___} — a site occupied by a south-facing walker

{4, a___} — a site occupied by a west-facing walker

where the quantity a___ indicates a sequence of zero or more values (e.g., numbers, symbols, or lists) representing the characteristics of the individual (see Appendix A for a detailed discussion of the use of underscores in pattern-matching).

At each time step, a walker, regardless of whether he moves or remains in place, randomly chooses a direction to face, using

```
RND := Random[Integer, {1, 4}]
```

The following 28 update rules are used to move individuals.

A walker facing an empty site moves from the site he is occupying unless one or more other walkers face the same empty site in which case he remains in place and randomly chooses a direction to face.

```
walk[{1,a___},0,_,_,_,{4,___},_,_,_,_,_,_,_] := {RND,a}
walk[{1,a___},0,_,_,_,_,_,_,{2,___},_,_,_,_] := {RND,a}
walk[{1,a___},0,_,_,_,_,_,_,{3,___},_,_,_] := {RND,a}
walk[{1,a___},0,_,_,_,_,_,_,_,_,_,_,_] := 0
walk[{2,a___},_,0,_,_,{3,___},_,_,_,_,_,_] := {RND,a}
walk[{2,a___},_,0,_,_,_,{1,___},_,_,_,_,_] := {RND,a}
walk[{2,a___},_,0,_,_,_,_,_,_,{4,___},_,_] := {RND,a}
walk[{2,a___},_,0,_,_,_,_,_,_,_,_,_,_] := 0
walk[{3,a___},_,_,0,_,_,{4,___},_,_,_,_,_] := {RND,a}
walk[{3,a___},_,_,0,_,_,_,{2,___},_,_,_,_] := {RND,a}
walk[{3,a___},_,_,0,_,_,_,_,_,_,{1,___},_] := {RND,a}
walk[{3,a___},_,_,0,_,_,_,_,_,_,_,_] := 0
walk[{4,a___},_,_ ,_,0,_,_,{1,___},_,_,_,_,_] := {RND,a}
walk[{4,a___},_,_ ,_,0,_,_,_,{3,___},_,_,_,_] := {RND,a}
walk[{4,a___},_,_ ,_,0,_,_,_,_,_,_,{2,___}] := {RND,a}
walk[{4,a___},_,_ ,_,0,_,_,_,_,_,_,_,_] := 0
```

Any other walker remains in place and randomly selects a direction to face.

```
walk[{_,a___},_,_,_,_,_,_,_,_,_,_,_,_] := {RND,a}
```

An empty site remains empty if two or more walkers face it.

```
walk[0,{3,___},{4,___},_,_,_,_,_,_,_,_,_,_] := 0
walk[0,{3,___},_,{1,___},_,_,_,_,_,_,_,_,_] := 0
walk[0,{3,___},_,_,{2,___},_,_,_,_,_,_,_,_] := 0
walk[0,_,{4,___},{1,___},_,_,_,_,_,_,_,_,_] := 0
```

```
walk[0,_,{4,___},_,{2,___},_,_,_,_,_,_,_,_] := 0 .
walk[0,_,_,{1,___},{2,___},_,_,_,_,_,_,_,_] := 0
```

An empty site becomes occupied if it is faced by exactly one walker.

```
walk[0,{3,a___},_,_,_,_,_,_,_,_,_,_,_,_] := {RND,a}
walk[0,_,{4,a___},_,_,_,_,_,_,_,_,_,_,_] := {RND,a}
walk[0,_,_,{1,a___},_,_,_,_,_,_,_,_,_,_] := {RND,a}
walk[0,_,_,_,{2,a___},_,_,_,_,_,_,_,_,_] := {RND,a}
```

Any other empty site remains empty

```
walk[0,_,_,_,_,_,_,_,_,_,_,_,_] := 0
```

The walk rules are applied to the lattice sites using the anonymous function (anonymous functions are explained in Appendix A):

```
GN[walk, #]&
```

where

```
GN[func_, lat_] :=
  MapThread[func, Map[RotateRight[lat, #]&,
            {{0, 0}, {1, 0}, {0, -1}, {-1, 0}, {0, 1},
             {1, -1}, {-1, -1}, {-1, 1}, {1, 1}, {2, 0},
             {0, -2}, {-2, 0}, {0, 2}}], 2]
```

(the MapThread function is explained and its use is illustrated in Appendix B).

Interacting

There are two general categories of interaction that we consider.

Person-to-Person: The first mode of interaction is bilateral and occurs between two individuals facing one another on adjacent sites (e.g., a south (east, north, west) facing individual who is face-to-face with a north (west, south, east) facing individual on the adjacent southern (eastern, northern, western) site. This situation is illustrated as follows.

Two Facing Individuals Interacting with Each Other

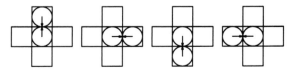

As can be seen from the preceding diagrams, rules for updating lattice sites for pair-wise interactions between facing individuals depend on the values of a site and its four neighboring sites (constituting what is called the vonNeumann neighborhood of a site). Accordingly, the rules for updating a lattice site take five arguments.

```
personToPersonInteract[site, N, E, S, W]
```

where the five arguments represent the value of the site and the values of the four nearest neighbors in the N, E, S, W directions.

The interaction rules for person-to-person interactions are applied to the lattice using the anonymous function

```
vonNeumann[personToPersonInteract, #]&
```

where

```
vonNeumann[func_, lat_] :=
  MapThread[func, Map[RotateRight[lat, #]&,
        {{0, 0}, {1, 0}, {0, -1}, {-1, 0}, {0, 1}}], 2]
```

Person-to-Group: The second mode of interaction is multilateral and involves an individual interacting with two or more of the individuals on adjacent nearest neighbor sites. This situation is illustrated as follows.

An Individual Interacting with His Neighbors

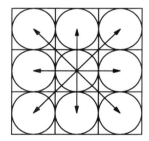

Note: All of the sites in the neighborhood shown previously are depicted as being occupied by individuals. In most of the cases considered in this book, one or more of these sites will be empty (except when the lattice is completely occupied by people).

As can be seen from the preceding neighborhood diagram, updating the lattice sites for person-to-group interactions depends on the values of the site and its eight neighboring sites (constituting what is called the Moore neighborhood of a site). Accordingly, the rules for updating a lattice site take nine arguments.

```
personToGroupInteract[site, N, E, S, W, NE, SE, SW, NW]
```

where the nine arguments represent the value of the site, the values of the four nearest neighbors in the N, E, S, W directions, and the values of the four nearest neighbors in the NE, SE, SW, NW directions.

The interaction rules for person-to-group interactions are applied to the lattice using

```
Moore[personToGroupInteract, #]&
```

where

```
Moore[func_, lat_] :=
  MapThread[func, Map[RotateRight[lat, #]&,
            {{0, 0}, {1, 0}, {0, -1}, {-1, 0}, {0, 1},
             {1, -1}, {-1, -1}, {-1, 1}, {1, 1}}], 2]
```

Finally, we combine the partial steps that occur during a time step by creating a nested anonymous function. This is illustrated for the cases

```
GN[walk, vonNeumann[personToPersonInteract, #]]&

GN[walk, Moore[personToGroupInteract, #]]&

GN[walk, vonNeumann[personToPersonInteract,
                 Moore[personToGroupInteract, #]]]&
```

Note: The first two cases involve a time step in which movement and either person-to-person or person-to-group interactions occur, and the last case shows a hypothetical case in which both person-to-person and person-to-group interactions occur (an example of this might be a model in which an individual's behavior when interacting with another individual depends on the attitudes of the other people in the neighborhood towards that other individual).

1.2.4 Evolving the System

The system evolves over t time steps, starting with the initial lattice configuration, society, using one of the following nesting operations (the NestList function is explained and examples of its use are shown in Appendix B).

```
NestList[GN[walk,
      vonNeumann[personToPersonInteract, #]]&, society, t]

NestList[GN[walk,
            Moore[personToGroupInteract, #]]&, society, t]

NestList[GN[walk, vonNeumann[personToPersonInteract,
            Moore[personToGroupInteract, #]]]&, society, t]
```

1.2.5 The Program

Putting all of the code fragments together, we can write the following program for a system of mobile, interacting heterogeneous individuals. We assume that no interaction takes place between individuals here in order to keep the program simple since the nature of the interact rules will vary from model to model.

n = size of lattice
p = population density
t = number of time steps

```
randomWalkers[n_, p_, g_, m_, s_, r_, t_] :=
Module[{society, RND, k, walk, GN},
  RND:= Random[Integer, {1, 4}];
  k = 0;
  society = Table[Floor[p + Random[]], {n}, {n}] /.
    1 :> {RND, ++k, Floor[g + Random[]],
        Table[Random[Integer, {1, m}], {s}],
        Random[Integer, {0, r}]};

  walk[{1,a___},0,_,_,_,{4,___},_,_,_,_,_,_] := {RND,a};
  walk[{1,a___},0,_,_,_,_,_,_,{2,___},_,_,_,_] := {RND,a};
  walk[{1,a___},0,_,_,_,_,_,_,_,{3,___},_,_,_] := {RND,a};
  walk[{1,a___},0,_,_,_,_,_,_,_,_,_,_,_] := 0;
  walk[{2,a___},_,0,_,_,{3,___},_,_,_,_,_,_] := {RND,a};
  walk[{2,a___},_,0,_,_,_,{1,___},_,_,_,_,_] := {RND,a};
  walk[{2,a___},_,0,_,_,_,_,_,_,{4,___},_,_] := {RND,a};
  walk[{2,a___},_,0,_,_,_,_,_,_,_,_,_] := 0;
  walk[{3,a___},_,_,0,_,_,{4,___},_,_,_,_,_] := {RND,a};
  walk[{3,a___},_,_,0,_,_,_,{2,___},_,_,_,_] := {RND,a};
  walk[{3,a___},_,_,0,_,_,_,_,_,_,{1,___},_] := {RND,a};
  walk[{3,a___},_,_,0,_,_,_,_,_,_,_,_] := 0;
  walk[{4,a___},_,_,_,0,_,_,{1,___},_,_,_,_] := {RND,a};
  walk[{4,a___},_,_,_,0,_,_,_,{3,___},_,_,_] := {RND,a};
  walk[{4,a___},_,_,_,0,_,_,_,_,_,_,{2,___}] := {RND,a};
  walk[{4,a___},_,_,_,0,_,_,_,_,_,_,_] := 0;
  walk[{_,a___},_,_,_,_,_,_,_,_,_,_,_,_] := {RND,a};
  walk[0,{3,___},{4,___},_,_,_,_,_,_,_,_,_,_] := 0;
  walk[0,{3,___},_,{1,___},_,_,_,_,_,_,_,_,_] := 0;
  walk[0,{3,___},_,_,{2,___},_,_,_,_,_,_,_,_] := 0;
  walk[0,_,{4,___},{1,___},_,_,_,_,_,_,_,_,_] := 0;
  walk[0,_,{4,___},_,{2,___},_,_,_,_,_,_,_,_] := 0;
  walk[0,_,_,{1,___},{2,___},_,_,_,_,_,_,_,_] := 0;
  walk[0,{3,a___},_,_,_,_,_,_,_,_,_,_,_] := {RND,a};
  walk[0,_,{4,a___},_,_,_,_,_,_,_,_,_,_] := {RND,a};
  walk[0,_,_,{1,a___},_,_,_,_,_,_,_,_,_] := {RND,a};
  walk[0,_,_,_,{2,a___},_,_,_,_,_,_,_,_] := {RND,a};
  walk[0,_,_,_,_,_,_,_,_,_,_,_,_] := 0;

  GN[func_, lat_] :=
    MapThread[func, Map[RotateRight[lat, #]&,
        {{0, 0}, {1, 0}, {0, -1}, {-1, 0}, {0, 1},
         {1, -1}, {-1, -1}, {-1, 1}, {1, 1}, {2, 0},
         {0, -2}, {-2, 0}, {0, 2}}], 2];

  NestList[GN[walk, #]&, society, t]]
```

1.2.6 Running the Simulation

We run the randomWalkers program on a 100 by 100 lattice having a 65% population of individuals, over 10,000 time steps.

```
SeedRandom[2];
results = randomWalkers[100, 0.65, 1, 1, 1, 1, 10000];
```

Note: The use of the SeedRandom function (which is explained in Appendix B) allows the same list of random numbers to be generated over and over again.

The initial and final configurations of the lattice are shown in the following graphic, in which the shading of an individual is determined by his or her initial location on the lattice and an empty site is shown as white.

```
m = Max[First[results]];
Show[GraphicsArray[
  Map[Graphics[RasterArray[
    Map[If[# === 0, GrayLevel[1], GrayLevel[#[[2]]/m]]&,
                    #, {2}]], AspectRatio → Automatic]&,
  {First[results], Last[results]}]]]];
```

The initially smooth gradation in shading from the top to the bottom of the lattice over a long time turns into a homogeneous mixture as a thorough mixing of the population occurs via the random movement of the individuals in the system.

1.3 Computer Simulation Projects

1. Create a lattice configuration in which all of the sites occupied by individuals are initially clustered together in the middle of the lattice. Use this lattice configuration for society in the randomWalkers model, run the resulting program and observe the dispersion of people from the initial cluster into the surrounding areas.

2. In the course of moving around the lattice, individuals will encounter other individuals face-to-face on adjacent sites. In many of the models in the following chapters, there bilateral interactions such as economic transactions, information sharing, or cultural exchanges between these individuals. The randomWalkers program can be easily modified to allow individuals to keep track of the number of individuals they encounter in the course of their travels. This can be done by adding an element to the list of attributes of an individual to serve as a counter of the number of face-to-face encounters that he has over time. The initial value of this element should be zero and the value should increase by one every time the individual finds himself face-to-face with another individual. The counting process can be carried out in either of two ways: by adding four rules to the walk rule set to cover the face-to-face situation, or by writing five personToPersonInteract rules, four to cover the four situations in which an individual encounters another person face-to-face and one to cover all other possible situations. The first method has the benefit of producing more concise code that runs somewhat faster whereas the second method has the benefit of making the code easier to understand. Implement both of these counting methods in variants of the randomWalkers program and then run these programs. Calculate the average number of encounters that an individual has as a function of time.

3. We can allow an individual to keep a record of whom he encounters and how often he encounters them. This will be useful for some of the phenomena we discuss in the chapters that follow. This can be done by using as an additional element in an individual's attribute list, a list that is initially empty and to which a person's name is added whenever the individual encounters that person face-to-face. Modify the code you wrote in the previous project to allow this to be done. Run a program with these modified rules and calculate the average number of multiple encounters an individual has with other individuals as a function of time.

4. Using the program you developed in the previous project, determine the average number of new people who are encountered over time by an individual as a function of the population density. This can be done by eliminating duplicates in the encounter lists in the previous project. Look to see if this quantity passes through a maximum as a function of population density and explain why there is or is not a maximum.

5. It is important to know if the results obtained in the simulations in the previous three projects are sensitive to the configuration of the system at the start of the run.

The easiest way to check this would be to run the randomWalkers program several times using a different SeedRandom value (the SeedRandom function is explained in Appendix B) for each run. The problem with this method is that it's too *shotgun* in nature, resulting in a large number of the people in the system having different starting locations and orientations. An alternative, more delicate, method would be to modify the program so that after everyone has been placed on the lattice a small number of them are altered in some way (e.g., by changing their orientation or by exchanging their locations). Then the original and altered programs can be run with the same SeedRandom value and the results can be compared. The code modification can be done using one or more simple transformation rule (the Rule and ReplaceAll functions are explained in Appendix A) placed immediately after the society definition. Implement this and check to see if there is an initial condition sensitivity and if there is, how long it persists during the course of a simulation run.

1.4 References

Gaylord, R.J. and K. Nishidate. 1996. "Cellular Automaton Model for Random Walkers." *Physical Review Letters* 77: 1675–1678.

Gaylord, R. J. and K. Nishidate. 1996. *MODELING NATURE: Cellular Automata Simulations Using Mathematica.* New York: TELOS/Springer-Verlag. Japanese edition, 1997.

1.5 Programs in the Chapter

1.5.1 randomWalkers

```
randomWalkers[n_, p_, g_, m_, s_, r_, t_] :=
Module[{society, RND, k, walk, GN},
  RND:= Random[Integer, {1, 4}];
  k = 0;
  society = Table[Floor[p + Random[]], {n}, {n}] /.
    1 :> {RND, ++k, Floor[g + Random[]],
      Table[Random[Integer, {1, m}], {s}],
      Random[Integer, {0, r}]};
  walk[{1,a___},0,_,_,_,{4,___},_,_,_,_,_,_,_] := {RND,a};
  walk[{1,a___},0,_,_,_,_,_,_,{2,___},_,_,_,_] := {RND,a};
  walk[{1,a___},0,_,_,_,_,_,_,_,{3,___},_,_,_] := {RND,a};
  walk[{1,a___},0,_,_,_,_,_,_,_,_,_,_,_] := 0;
  walk[{2,a___},_,0,_,_,{3,___},_,_,_,_,_,_,_] := {RND,a};
  walk[{2,a___},_,0,_,_,_,{1,___},_,_,_,_,_,_] := {RND,a};
  walk[{2,a___},_,0,_,_,_,_,_,_,{4,___},_,_] := {RND,a};
  walk[{2,a___},_,0,_,_,_,_,_,_,_,_,_,_] := 0;
  walk[{3,a___},_,_,0,_,_,{4,___},_,_,_,_,_,_] := {RND,a};
  walk[{3,a___},_,_,0,_,_,_,{2,___},_,_,_,_,_] := {RND,a};
  walk[{3,a___},_,_,0,_,_,_,_,_,_,{1,___},_] := {RND,a};
  walk[{3,a___},_,_,0,_,_,_,_,_,_,_,_,_] := 0;
  walk[{4,a___},_,_,_,0,_,_,{1,___},_,_,_,_,_] := {RND,a};
  walk[{4,a___},_,_,_,0,_,_,_,{3,___},_,_,_,_] := {RND,a};
  walk[{4,a___},_,_,_,0,_,_,_,_,_,_,{2,___}] := {RND,a};
  walk[{4,a___},_,_,_,0,_,_,_,_,_,_,_,_] := 0;
  walk[{_,a___},_,_,_,_,_,_,_,_,_,_,_,_] := {RND,a};
  walk[0,{3,___},{4,___},_,_,_,_,_,_,_,_,_,_] := 0;
  walk[0,{3,___},_,{1,___},_,_,_,_,_,_,_,_,_] := 0;
  walk[0,{3,___},_,_,{2,___},_,_,_,_,_,_,_,_] := 0;
  walk[0,_,{4,___},{1,___},_,_,_,_,_,_,_,_,_] := 0;
  walk[0,_,{4,___},_,{2,___},_,_,_,_,_,_,_,_] := 0;
  walk[0,_,_,{1,___},{2,___},_,_,_,_,_,_,_,_] := 0;
  walk[0,{3,a___},_,_,_,_,_,_,_,_,_,_,_] := {RND,a};
  walk[0,_,{4,a___},_,_,_,_,_,_,_,_,_,_] := {RND,a};
  walk[0,_,_,{1,a___},_,_,_,_,_,_,_,_,_] := {RND,a};
  walk[0,_,_,_,{2,a___},_,_,_,_,_,_,_,_] := {RND,a};
  walk[0,_,_,_,_,_,_,_,_,_,_,_,_] := 0;
  GN[func_, lat_] :=
    MapThread[func, Map[RotateRight[lat, #]&,
        {{0, 0}, {1, 0}, {0, -1}, {-1, 0}, {0, 1},
         {1, -1}, {-1, -1}, {-1, 1}, {1, 1}, {2, 0},
         {0, -2}, {-2, 0}, {0, 2}}], 2];
  NestList[GN[walk, #]&, society, t]]
```

Cultural Exchange

"I know of no more valuable commodity than information."

—Michael Douglas in *Wall Street* (1987)

2

Transmitting Culture

"Role models are important."

—Peter Weller in *Robocop*

2.1 Introduction

How do people come to have shared values based on ideas, beliefs, likes and dislikes, or attitudes? One possible mechanism for the spreading of values through a population is through a sort of contagious process, occurring as individuals come into contact with one another and interact. This interaction results in a form of imitative behavior sometimes referred to as *cultural transmission* or *social learning*. In this chapter, we look at several models of the change of values in a mobile society. In later chapters, we use some of the ideas employed in these cultural transmission models to examine various other socioeconomic behaviors.

2.2 The More Alike We Are, the More Alike We Become

Axelrod [1997] has proposed a model for the transmission of culture. Using a square lattice, each lattice site is occupied by an *agent* (known as a homogeneous village) and the lattice has absorbing boundaries so villages on the interior of the lattice have four nearest neighbors (to the north, east, south, and west), villages on the edge of the lattice have three nearest neighbors, and corner villages have two nearest neighbors. A village is characterized by having five attributes (called *features*), each of which has an integer value (called a *trait*) ranging from 1 to 10. The system evolves over a number of time steps.

At each time step, one village (called the active village) is randomly chosen. This active village randomly selects a village on a nearest neighbor site. The active village interacts with the selected village as follows: a comparison is made between the traits of corresponding features of the two villages. If the traits of each corresponding feature are the same (e.g., {5, 9, 1, 3, 2} and {5, 9, 1, 3, 2}), nothing happens. If any of the traits of the features of the active village differ from the trait of the corresponding feature of the selected village, then a *cultural interaction* occurs with a probability equal to the fraction of features that share the same trait, that is, to their degree of *cultural similarity*. For example, if the traits of the active village's features are {3, 2, 1, 7, 5} and the traits of the selected village's features are {4, 8, 1, 2, 5}, then there is 40% probability that the villages will interact culturally because the traits of the third and fifth features of the two villages are identical. If a cultural interaction does occur, it is carried out by randomly choosing one of the features of the active village which has a different trait from the trait of the corresponding feature of the selected village and changing the active village's feature trait to that of the trait of the selected village's corresponding feature. Thus, if the active village having features {3, 2, 1, 7, 5} does interact with the selected village having features {4, 8, 1, 2, 5}, the interaction will be carried out by randomly choosing the first, second, or fourth feature of the active village and changing its trait to the trait of the corresponding feature of the selected village, so that if, for example, the second feature is chosen, the active village ends up with feature traits of {3, 8, 1, 7, 5}. This process continues until the system becomes invariant.

The Axelrod model is quite elegant in its simplicity (a fundamental rule of scientific model building is to follow the dictums of both the military, "keep it simple, stupid" (the KISS principle), and Albert Einstein, "a [model] should be as simple as possible, but no simpler"). We can make two modifications to the model that will extend its usefulness, while maintaining its essential simplicity.

• Incorporating mobility.

Axelrod states that "since there is no movement in the model, the sites themselves can be thought of as homogenous villages." If some of the lattice sites were empty and other sites were occupied by individuals, the model could be applied to a mobile society using the walk rules developed in Chapter 1.

• Incorporating bilateral cultural exchange.

The pair-wise interaction between villages in the Axelrod model is one-way or unilateral in that a randomly chosen active village changes the trait of its features as a result of its interaction with a selected nearest neighbor village and the selected village is left unchanged.

In our variant of the Axelrod model, individuals roam around a lattice, carrying with them cultural attributes, and altering these attributes as they engage in bilateral interactions with individuals they encounter along the way.

2.2.1 The System

Our model uses an n by n square lattice with wraparound boundary conditions. There is a population density p of individuals occupying lattice sites and the remaining sites are empty. Each person is characterized by the direction he is facing and a *meme* list with s elements. The system evolves over a given number of time steps or until all of the agents have identical meme lists.

Note: The term *meme* was coined by Richard Dawkins to represent the basic unit of cultural transmission, analogous to the gene which is the basic unit of genetic transmission.

2.2.2 Populating Society

We first need to specify the values of the lattice sites.

- An empty site has value 0.

- A site that is occupied by an individual has a value of an ordered pair in which the first component is a randomly chosen integer between 1 and 4 (indicating that the individual is facing north, east, south, or west, respectively), and the second component is a meme list of s elements, each of which has an integer value between 1 and m (giving each of the s memes of an individual a measure).

The code for creating this system is given by:

```
RND := Random[Integer, {1, 4}]
society = Table[Floor[p + Random[]], {n}, {n}] /.
          1 :> {RND, Table[Random[Integer, {1, m}], {s}]}
```

2.2.3 Executing a Time Step

The time step will be executed in three consecutive *partial-steps*, the first two involving the cultural exchange between individuals facing one another on adjacent sites and the last one involving the movement of individuals.

Deciding

Any individual who is facing another individual on an adjacent site (e.g., a south (east, north, west) facing individual who is face-to-face with a north (west, south, east) facing individual on the adjacent site) interacts culturally with that individual. The cultural exchange that occurs between facing individuals follows the Axelrod model, except that it is a two-way interaction between the two individuals, allowing both to be changed by the interaction. A comparison is made between the values of corresponding memes of the two individuals. If all of the values of the corresponding memes are the same for each individual, nothing happens. If any of the values of the corresponding memes of the two individuals differ, then cultural transmission or *social learning* occurs with a probability equal to the percentage of memes that *share* the same value. If cultural transmission does occur, it is carried out by randomly choosing one of the memes whose value differs for the two individuals and changing that value for *both* individuals.

For example, if the south-facing individual's meme list is {3, 2, 1, 7, 5} and its north-facing neighbor's meme list is {4, 8, 1, 2, 5}, there is a 60% probability that the meme list will remain unchanged and a 40% probability that the value of the first, second, or fourth meme of each individual will be changed to an integer value somewhere between the values of that meme for the two individuals (thus, if the second meme were randomly chosen, the individuals' meme lists would become {3, x, 1, 7, 5} and {4, x, 1, 2, 5}, where x is a randomly chosen integer between 2 and 8.

We first let individuals facing one another on adjacent sites decide which meme to change and what the meme's value should be. The actual change will take place on the next partial-step. As in a fair coin toss where it does not matter who flips the coin and who makes the call of heads or tails, it is irrelevant which of the two interacting individuals determines the cultural exchange for both individuals. We let the south (east) facing individual decide for both himself and his north (west) facing neighbor.

The lattice sites are updated using the following rules.

```
decide[{3, y_}, _, _, {1, y_}, _] := {3, y}
decide[{2, y_}, _, {4, y_}, _, _] := {2, y}
decide[{3, y_}, _, _, {1, z_}, _] := Module[
   {diffs = Flatten[Position[Abs[y - z], _?Positive]]},
   {3, y, Random[Integer, {y[[#]], z[[#]]}], #}&[
        diffs[[Random[Integer, {1, Length[diffs]}]]]]] /;
                Random[Integer, {1,s}] > Length[diffs]]
decide[{2, y_}, _, {4, z_}, _, _] := Module[
   {diffs = Flatten[Position[Abs[y - z], _?Positive]]},
   {2, y, Random[Integer, {y[[#]], z[[#]]}], #}&[
        diffs[[Random[Integer, {1, Length[diffs]}]]]]] /;
                Random[Integer, {1,s}] > Length[diffs]]
decide[a_, _, _, _, _] := a
```

The first two decide rules apply to a south (east) facing individual having a north (west) facing individual on the adjacent southern (eastern) site, when the meme lists of both are identical. In this case there is no change in meme list values.

The third and fourth decide rules apply to a south (east) facing individual having a north (west) facing individual on the adjacent southern (eastern) site, when the meme lists of the two are not identical.

The calculation involved in the execution of these two rules can be illustrated using the two meme lists, {3, 2, 1, 7, 5} and {4, 8, 1, 2, 5}. The following calculations occur.

- For the meme lists, y and z, a list, diffs = Flatten[Position[Abs[y - z], _?Positive]] = {1, 2, 4}, is calculated whose elements are the positions in the meme lists where the meme values differ.

- Using diffs[[Random[Integer, {1, Length[diffs]}]]], one of the locations where the meme values differ is randomly chosen. For {1, 2, 4}[[Random[Integer, {1, Length[{1, 2, 4}]}]]], the third component in diffs, 4, representing the fourth meme in the meme list, might be chosen.

- Using 4 as # in the anonymous function {3, y, Random[Integer, {y[[#]], z[[#]]}], #}& returns {3, y, Random[Integer, {y[[4]], z[[4]]}], 4}.

- The quantity Random[Integer, {y[[4]], z[[4]]}] returns a random integer between the values of the two memes at the selected location. In our example, we have Random[Integer, {7, 2}] which might give 6.

- The computed values (6 and 4) are appended, resulting in {3, {3, 2, 1, 7, 5}, 6, 4}.

The third and fourth decide rules are applied with a probability equal to the number of memes that share the same value. To do this, the condition Random[Integer, {1, s}] > Length[diffs] is used. For our example, we would have Random[Integer, {1, 5}] > 3 so that for facing individuals with nonidentical meme lists, these rules would be applied 40% of the time. The rest of the time, the individuals would follow the fifth decide rule.

The fifth decide rule applies to all other sites (either empty or occupied by an individual) and leaves them unchanged.

The decide rules are applied to the lattice sites using

```
vonNeumann[decide, #]&
```

where

```
vonNeumann[func_, lat_] :=
  MapThread[func, Map[RotateRight[lat, #]&,
          {{0, 0}, {1, 0}, {0, -1}, {-1, 0}, {0, 1}}], 2]
```

Exchanging Meme Values

Having decided the nature of the cultural exchange, we can carry out the exchange, using the rules:

```
exchange[{3, y_, c_, d_}, _, _, _, _] :=
                            {3, ReplacePart[y, c, d]}
exchange[{2, y_, c_, d_}, _, _, _, _] :=
                            {2, ReplacePart[y, c, d]}
exchange[{1, z_}, {3, y_, c_, d_}, _, _, _] :=
                            {1, ReplacePart[z, c, d]}
exchange[{4, z_}, _, _, _, {2, y_, c_, d_}] :=
                            {4, ReplacePart[z, c, d]}
exchange[a_, _, _, _, _] := a
```

In the preceding exchange rules, the individuals with four-element lists are the ones who have made the decision on what meme to change and what the new meme value should be. The first two exchange rules use these values, which are given by the third and fourth list elements, to make the change in these persons. The next two exchange rules use these same values to make the changes in the meme lists of the people facing the decision makers. The fifth rule applies to people who do not exchange values and to empty sites.

The exchange rules are applied to the lattice sites using the anonymous function

```
vonNeumann[exchange, vonNeumann[decide, #]]&
```

Moving

The 28 rules for the movement of individuals are given by the set of walk rules in Chapter 1.

The walk rules are applied to the lattice sites using the anonymous function

```
GN[walk, vonNeumann[exchange, vonNeumann[decide, #]]]&
```

where

```
GN[func_, lat_] :=
  MapThread[func, Map[RotateRight[lat, #]&,
          {{0, 0}, {1, 0}, {0, -1}, {-1, 0}, {0, 1},
           {1, -1}, {-1, -1}, {-1, 1}, {1, 1}, {2, 0},
           {0, -2}, {-2, 0}, {0, 2}}], 2]
```

2.2.4 Evolving the System

The system evolves over t time steps, starting with the initial lattice configuration, society, using the following nesting operation.

```
NestList[GN[walk, vonNeumann[exchange,
                    vonNeumann[decide, #]]]&, society, t]
```

2.2.5 The Program

n = size of lattice
p = population density
s = size of meme list (number of memes)
m = maximum cultural value (values from 1 to m)
t = number of time steps

```
cultureSpreadingShared[n_, p_, s_, m_, t_] :=
Module[{society, RND, decide, exchange,
                    walk, vonNeumann, GN},

  RND:= Random[Integer, {1, 4}];
  society = Table[Floor[p + Random[]], {n}, {n}] /.
        1 :> {RND, Table[Random[Integer, {1, m}], {s}]};

  decide[{3, y_}, _, _, {1, y_}, _] := {3, y};
  decide[{2, y_}, _, {4, y_}, _, _] := {2, y};
  decide[{3, y_}, _, _, {1, z_}, _] := Module[
    {diffs = Flatten[Position[Abs[y - z], _?Positive]]},
    {3, y, Random[Integer, {y[[#]], z[[#]]}], #}&[
        diffs[[Random[Integer, {1, Length[diffs]}]]]] /;
                Random[Integer, {1,s}] > Length[diffs]];
  decide[{2, y_}, _, {4, z_}, _, _] := Module[
    {diffs = Flatten[Position[Abs[y - z], _?Positive]]},
    {2, y, Random[Integer, {y[[#]], z[[#]]}], #}&[
        diffs[[Random[Integer, {1, Length[diffs]}]]]] /;
                Random[Integer, {1,s}] > Length[diffs]];
  decide[a_, _, _, _, _] := a;

  exchange[{3, y_, c_, d_}, _, _, _, _] :=
                        {3, ReplacePart[y, c, d]};
  exchange[{2, y_, c_, d_}, _, _, _, _] :=
                        {2, ReplacePart[y, c, d]};
  exchange[{1, z_}, {3, y_, c_, d_}, _, _, _] :=
                        {1, ReplacePart[z, c, d]};
  exchange[{4, z_}, _, _, _, {2, y_, c_, d_}] :=
                        {4, ReplacePart[z, c, d]};
  exchange[a_, _, _, _, _] := a;

  (* walk rules go here *)

  vonNeumann[func_, lat_] :=
    MapThread[func, Map[RotateRight[lat, #]&,
        {{0, 0}, {1, 0}, {0, -1}, {-1, 0}, {0, 1}}], 2];
```

```
GN[func_, lat_] :=
  MapThread[func, Map[RotateRight[lat, #]&,
        {{0, 0}, {1, 0}, {0, -1}, {-1, 0}, {0, 1},
          {1, -1}, {-1, -1}, {-1, 1}, {1, 1}, {2, 0},
          {0, -2}, {-2, 0}, {0, 2}}], 2];

NestList[GN[walk, vonNeumann[exchange,
              vonNeumann[decide, #]]]&, society, t]]
```

2.2.6 Running the Simulation

We run the cultureSpreadingShared program on a 25 by 25 lattice having a 70% population density of individuals, each of whom can have two possible values for each of its two memes, over 500 time steps.

```
SeedRandom[3];
results = cultureSpreadingShared[25, 0.7, 2, 2, 500];
```

The displayCulture function graphically depicts a lattice configuration, using different colors to represent different meme configurations and drawing empty sites in gray.

```
displayCulture[lat_] :=
  Show[Graphics[RasterArray[lat /.
      {0 → RGBColor[0.5, 0.5, 0.5],
       {_Integer, {n_, m_}} :→ RGBColor[n - 1, m - 1, 0]}]],
    AspectRatio → Automatic];
```

Applying the displayCulture program to the lattice configurations generated by the program run at various times,

```
Show[GraphicsArray[
  {{displayCulture[results[[1]]],
    displayCulture[results[[167]]]},
   {displayCulture[results[[333]]],
    displayCulture[results[[500]]]}}]];
```

Another way to visualize the changes in the cultural composition of the society over time is to draw the area between two adjacent individuals in white if their meme lists are identical, and in darker shades as the differences in meme list values increase. We can illustrate this using the simulation run that produced the preceding raster array graphic and showing the initial and final lattice configurations.

```
similarities[0, __] := {1, 1, 1, 1};
similarities[x_, r__] :=  Map[Count[#, 0]&,
    Map[Last[x] - Last[#]&, {r} /. 0 :> x]] / 2.;

vonNeumann[func_, lat_] :=
  MapThread[func, Map[RotateRight[lat, #]&,
      {{0, 0}, {1, 0}, {0, -1}, {-1, 0}, {0, 1}}]], 2];
```

```
makeBlock[{a_, b_, c_, d_}, {x_, y_}] := {
  GrayLevel[a],
  Polygon[{{x, y}, {x, y + 1}, {x + .5, y + .5}}],
  GrayLevel[b],
  Polygon[{{x,  y + 1}, {x + 1, y + 1}, {x + .5, y + .5}}],
  GrayLevel[c],
  Polygon[{{x + 1, y}, {x + 1, y + 1}, {x + .5, y + .5}}],
  GrayLevel[d],
  Polygon[{{x,  y}, {x + 1, y}, {x + .5, y + .5}}]};

Show[GraphicsArray[{
  Graphics[MapIndexed[makeBlock,
      vonNeumann[similarities, results[[1]]], {2}]],
  Graphics[MapIndexed[makeBlock,
      vonNeumann[similarities, results[[-1]]], {2}]]}],
  AspectRatio -> Automatic];
```

Yet another way to see the convergence of shared values in a community over time is
to look at the number of individuals that share meme lists. This can be done for a
society of individuals having a meme list of 2 memes, each of which can have a value
of 1 or 2, using the following code to generate a percentile bar chart of the number of
individuals having a particular meme list, in this case {1, 1}, {1, 2}, {2, 1}, or {2, 2}.

```
<<Graphics`;
countCultures[lat_] := Map[Count[lat, {_, #}, 2]&,
                           {{1, 1}, {1, 2}, {2, 1}, {2, 2}}];
counts[res_] := Map[countCultures,
                               res[[Range[1, 500, 25]]]];
Apply[PercentileBarChart, Transpose[counts[results]]];
```

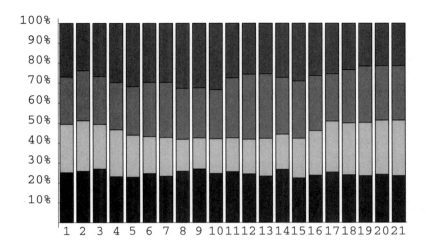

2.3 The Closer We Agree, the Closer Our Agreement Becomes

Another model of cultural transmission ignores how many meme values are shared between two facing individuals and instead randomly chooses a meme value to change and bases the change on how similar the chosen meme value is for the two individuals.

Deciding and Exchanging

The first two decide rules used in the previous model are used here. The third and fourth decide rules are changed so that each south (east) facing individual who is face-to-face with another individual on an adjacent site randomly selects a meme in his meme list to change and, with a probability equal to the *closeness* of the meme's value for the two individuals, determines what its new value will be by randomly choosing a new value between the meme values for the two individuals.

Each individual who is face-to-face with another individual on an adjacent site then makes the change in one of his meme values that was determined by the south (east) facing neighbor.

The set of decide rules used for this model are:

```
decide[{3, y_}, _, _, {1, y_}, _] := {3, y}
decide[{2, y_}, _, {4, y_}, _, _] := {2, y}
decide[{2, y_}, _, {4, z_}, _, _] :=
  Module[{r = Random[Integer, {1, s}]},
         {2, y, Random[Integer, {y[[r]], z[[r]]}], r} /;
         Random[Integer, {1, m}] > Abs[y[[r]] - z[[r]]]]]
```

```
decide[{3, y_}, _, _, {1, z_}, _] :=
  Module[{r = Random[Integer, {1, s}]},
          {3, y, Random[Integer, {y[[r]], z[[r]]}], r} /;
          Random[Integer,{1, m}] > Abs[y[[r]] - z[[r]]]]
decide[a_, _, _, _, _] := a
```

The exchange rules are the same as those used in the previous model.

2.3.1 The Program

```
cultureSpreadingCloseness[n_, p_, s_, m_, t_] :=
Module[{society, RND, decide, exchange,
                            walk, vonNeumann, GN},
  RND:= Random[Integer, {1, 4}];
  society = Table[Floor[p + Random[]], {n}, {n}] /.
          1 :> {RND, Table[Random[Integer, {1, m}], {s}]};

  decide[{3, y_}, _, _, {1, y_}, _] := {3, y};
  decide[{2, y_}, _, {4, y_}, _, _] := {2, y};
  decide[{2, y_}, _, {4, z_}, _, _] :=
    Module[{r = Random[Integer, {1, s}]},
          {2, y, Random[Integer, {y[[r]], z[[r]]}], r} /;
          Random[Integer,{1, m}] > Abs[y[[r]] - z[[r]]]];
  decide[{3, y_}, _, _, {1, z_}, _] :=
    Module[{r = Random[Integer, {1, s}]},
          {3, y, Random[Integer, {y[[r]], z[[r]]}], r} /;
          Random[Integer,{1, m}] > Abs[y[[r]] - z[[r]]]];
  decide[a_, _, _, _, _] := a;

  exchange[{3, y_, c_, d_}, _, _, _, _] :=
                            {3, ReplacePart[y, c, d]};
  exchange[{2, y_, c_, d_}, _, _, _, _] :=
                            {2, ReplacePart[y, c, d]};
  exchange[{1, z_}, {3, y_, c_, d_}, _, _, _] :=
                            {1, ReplacePart[z, c, d]};
  exchange[{4, z_}, _, _, _, {2, y_, c_, d_}] :=
                            {4, ReplacePart[z, c, d]};
  exchange[a_, _, _, _, _] := a;

  (* walk rules go here *)

  vonNeumann[func_, lat_] :=
    MapThread[func, Map[RotateRight[lat, #]&,
          {{0, 0}, {1, 0}, {0, -1}, {-1, 0}, {0, 1}}], 2];
```

```
GN[func_, lat_] :=
  MapThread[func, Map[RotateRight[lat, #]&,
          {{0, 0}, {1, 0}, {0, -1}, {-1, 0}, {0, 1},
           {1, -1}, {-1, -1}, {-1, 1}, {1, 1}, {2, 0},
           {0, -2}, {-2, 0}, {0, 2}}], 2];

NestList[GN[walk, vonNeumann[exchange,
                vonNeumann[decide, #]]]&, society, t]]
```

2.3.2 Running the Simulation

We run the cultureSpreadingCloseness program on a 25 by 25 lattice having a 70% population of individuals with two memes, each of which can have two possible values, over 500 time steps.

```
SeedRandom[5];
results = cultureSpreadingCloseness[25, 0.7, 2, 2, 500];
```

The initial and final cultural compositions of the society are shown in the following.

```
similarities[0, __] := {1, 1, 1, 1};
similarities[x_, r__] := Map[Count[#, 0]&,
      Map[Last[x] - Last[#]&, {r} /. 0 :> Last[x]]] / 2.;

vonNeumann[func_, lat_] :=
  MapThread[func, Map[RotateRight[lat, #]&,
          {{0, 0}, {1, 0}, {0, -1}, {-1, 0}, {0, 1}}], 2];

makeBlock[{a_, b_, c_, d_}, {x_, y_}] := {
  GrayLevel[a],
  Polygon[{{x, y}, {x, y + 1}, {x + .5, y + .5}}],
  GrayLevel[b],
  Polygon[{{x,  y + 1}, {x + 1, y + 1}, {x + .5, y + .5}}],
  GrayLevel[c],
  Polygon[{{x + 1, y}, {x + 1, y + 1}, {x + .5, y + .5}}],
  GrayLevel[d],
  Polygon[{{x,  y}, {x + 1, y}, {x + .5, y + .5}}]};

Show[GraphicsArray[{
  Graphics[MapIndexed[makeBlock,
        vonNeumann[similarities, results[[1]]], {2}]],
  Graphics[MapIndexed[makeBlock,
        vonNeumann[similarities, results[[-1]]], {2}]]}],
  AspectRatio -> Automatic];
```

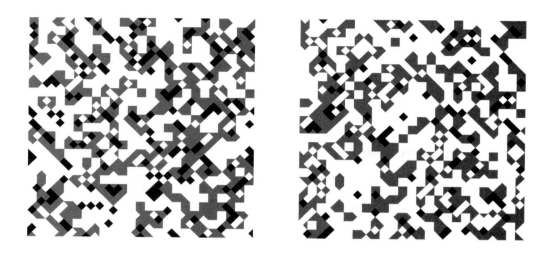

We can also look look at a percentile bar chart of the number of individuals with common meme lists.

```
<<Graphics`;
countCultures[lat_] := Map[Count[lat, {_, _, #}, 2]&,
                            {{1, 1}, {1, 2}, {2, 1}, {2, 2}}];
counts[res_] := Map[countCultures,
                        res[[Range[1, 500, 25]]]];
Apply[PercentileBarChart, Transpose[counts[results]]];
```

Comparing the figures obtained from the cultureSpreadingCloseness model to the figures obtained from the cultureSpreadingShared model, we see that the two models give similar results, indicating that there is no tendency for the population to adopt a common set of shared values. One way that convergence towards a shared value set might be achieved is to have a bias in the cultural exchange process, so that only one of the individuals changes one of his meme's values.

2.4 Social Status and Role Models

When two people of unequal social status interact culturally, it is more likely that the individual with the lower status will adopt a meme value of the individual with a higher social status while the meme value of the higher social status person will remain unchanged. This is most evident amongst young people who often adopt the attitude(s) of a so-called role model (hence, the famous advertising phrase "be like Mike"). In Chapter 5, we show how role models affect the spread of fads and fashions. We can model this effect using a simpler version of the previous two models because of the one-way nature of the cultural interaction. The necessary code changes follow.

Creating People with Varying Social Status

The value of a site that is occupied by an individual is a triplet of the form {direction, status, memelist}. The first component is a randomly chosen integer between 1 and 4, indicating that the individual is facing north, east, south, or west, respectively. The second component is a random real number between 0 and 1, representing the social status of the person. The third component is a meme list of s elements, each of which has an integer value between 1 and m.

Note: Although we use randomly chosen fixed social status values for each individual in this simple model, it is straightforward to use one or more changeable attributes of an individual, such as wealth, to determine social status. We can then change an individual's status over time as that attribute value changes (resource levels are discussed in Chapter 3). It is also possible to consider a model in which nonchangeable attributes, such as race or gender, play a role in determining status. Finally, we can combine these two by considering social status to be determined by some combination of changeable and unchangeable attributes, so that, for example, a changeable attribute that is important in determining the social status of people who are in a dominant or ruling position in a society, such as WASPS, becomes less important in determining the social status of people who are not in the ruling class, such as blacks and women.

The code for creating an n by n lattice with a population density p of individuals with varying social status is given by:

```
RND := Random[Integer, {1, 4}]
society = Table[Floor[p + Random[]], {n}, {n}] /. 1 :>
    {RND, Random[], Table[Random[Integer, {1, m}], {s}]}]
```

Deciding and Exchanging

No decide rules are needed in this model because the decision to change the value of a (randomly chosen) meme can be incorporated into the exchange rules by simply carrying out the exchange only if the person has a lower status than the individual he is facing. The exchange rules are given by:

```
exchange[{1, a_, u_}, {3, b_, v_}, _, _, _] :=
  {1, a, ReplacePart[u, v[[#]], #]&[
                    Random[Integer, {1, s}]]} /; a < b
exchange[{2, a_, u_}, _, {4, b_, v_}, _, _] :=
  {2, a, ReplacePart[u, v[[#]], #]&[
                    Random[Integer, {1, s}]]} /; a < b
exchange[{3, a_, u_}, _, _, {1, b_, v_}, _] :=
  {3, a, ReplacePart[u, v[[#]], #]&[
                    Random[Integer, {1, s}]]} /; a < b
exchange[{4, a_, u_}, _, _, _, {2, b_, v_}] :=
  {4, a, ReplacePart[u, v[[#]], #]&[
                    Random[Integer, {1, s}]]} /; a < b
```

All other individuals and empty sites remain unchanged.

```
exchange[z_, _, _, _, _] := z
```

2.4.1 The Program

```
charlesBarkley[n_, p_, s_, m_, t_] :=
Module[{society, RND, exchange, walk, vonNeumann, GN},

  RND := Random[Integer, {1, 4}];
  society = Table[Floor[p + Random[]], {n}, {n}] /. 1 :>
    {RND, Random[], Table[Random[Integer, {1, m}], {s}]};

  exchange[{1, a_, u_}, {3, b_, v_}, _, _, _] :=
    {1, a, ReplacePart[u, v[[#]], #]&[
                      Random[Integer, {1, s}]]} /; a < b;
  exchange[{2, a_, u_}, _, {4, b_, v_}, _, _] :=
    {2, a, ReplacePart[u, v[[#]], #]&[
                      Random[Integer, {1, s}]]} /; a < b;
  exchange[{3, a_, u_}, _, _, {1, b_, v_}, _] :=
    {3, a, ReplacePart[u, v[[#]], #]&[
                      Random[Integer, {1, s}]]} /; a < b;
  exchange[{4, a_, u_}, _, _, _, {2, b_, v_}] :=
    {4, a, ReplacePart[u, v[[#]], #]&[
                      Random[Integer, {1, s}]]} /; a < b;
  exchange[z_, _, _, _, _] := z;

  (* walk rules go here *)
```

```
vonNeumann[func_, lat_] :=
  MapThread[func, Map[RotateRight[lat, #]&,
        {{0, 0}, {1, 0}, {0, -1}, {-1, 0}, {0, 1}}], 2];

GN[func_, lat_] :=
  MapThread[func, Map[RotateRight[lat, #]&,
        {{0, 0}, {1, 0}, {0, -1}, {-1, 0}, {0, 1},
         {1, -1}, {-1, -1}, {-1, 1}, {1, 1}, {2, 0},
         {0, -2}, {-2, 0}, {0, 2}}], 2];

NestList[GN[walk, vonNeumann[exchange, #]]&, society, t]]
```

2.4.2 Running the Simulation

We run the charlesBarkley program on a 25 by 25 lattice having a 70% population of individuals with two memes, each of which can have two possible values, over 500 time steps, and look at the resulting graphics of the initial and final states of the system.

```
SeedRandom[7]
results = charlesBarkley[25, 0.7, 2, 2, 500];

similarities[0, __] := {1, 1, 1, 1};
similarities[x_, r__] := Map[Count[#, 0]&,
        Map[Last[x] - Last[#]&, {r} /. 0 :> Last[x]]] / 2.;

vonNeumann[func_, lat_] :=
  MapThread[func, Map[RotateRight[lat, #]&,
        {{0, 0}, {1, 0}, {0, -1}, {-1, 0}, {0, 1}}], 2];

makeBlock[{a_, b_, c_, d_}, {x_, y_}] := {
  GrayLevel[a],
  Polygon[{{x, y}, {x, y + 1}, {x + .5, y + .5}}],
  GrayLevel[b],
  Polygon[{{x,   y + 1}, {x + 1, y + 1}, {x + .5, y + .5}}],
  GrayLevel[c],
  Polygon[{{x + 1, y}, {x + 1, y + 1}, {x + .5, y + .5}}],
  GrayLevel[d],
  Polygon[{{x,   y}, {x + 1, y}, {x + .5, y + .5}}]};

Show[GraphicsArray[{
  Graphics[MapIndexed[makeBlock,
        vonNeumann[similarities, results[[1]]], {2}]],
  Graphics[MapIndexed[makeBlock,
        vonNeumann[similarities, results[[-1]]], {2}]]}],
  AspectRatio -> Automatic];
```

Comparing this figure to the corresponding figures for the cultureSpreadingShared and cultureSpreadingCloseness programs, we see that there are many fewer distinct regions in the charlesBarkley program.

We can also look at a percentile bar chart of the number of individuals with common meme lists.

```
<<Graphics`;
countCultures[lat_] := Map[Count[lat, {_, _, #}, 2]&,
                        {{1, 1}, {1, 2}, {2, 1}, {2, 2}}];
counts[res_] := Map[countCultures,
                              res[[Range[1, 500, 25]]]];
Apply[PercentileBarChart, Transpose[counts[results]]];
```

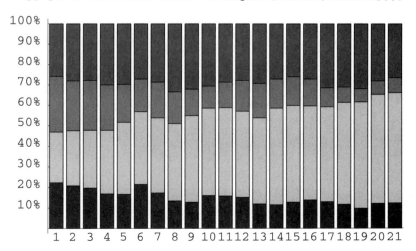

Comparing this figure to the corresponding figures for the cultureSpreadingShared and cultureSpreadingCloseness programs, we see that the percentage of individuals with the same set of meme value changes over time to a much greater extent in the charlesBarkley program.

Thus, the presence of individuals who can serve as role models for others dramatically increases the convergence of the society towards a common meme list and shared values. This is known as a social norm and is discussed in Chapter 5.

2.5 Computer Simulation Projects

1. Compare the results obtained from the progam for cultural transmission based on the closeness of a randomly chosen meme value and the program for cultural transmission based on the number of identical meme values.

2. In our model of the transmission of culture between people of differing social status, the social status is set for each individual at the start of the simulation and remains fixed thereafter. An interesting variation of this model would be to make social status endogenous. For example, the social status of a person could increase each time he interacts with a person of higher social status. Write several different variants of the charlesBarkley program, each with a different way in which the social status of an individual changes as he interacts with another person and compare the results obtained from running these simulations.

3. A number of interesting questions arise about localization and homogenization effects in contagious culture spreading:

- Is there is a tendency for spatial clusters of *like-minded* individuals to form?

- If regional clustering occurs, do these clusters eventually disappear at sufficiently long times or are they persistent?

- If clusters persist, do the values of individuals in different clusters eventually grow less alike, resulting in regional differences, or more alike, finally resulting in a single global meme set?

Note: To answer these questions, it would be useful to be able to identify clusters of individuals with shared values, where a cluster is a set of contiguous lattice sites that are populated by individuals whose meme lists are identical. We can easily determine the number and location of these clusters as follows.

We first assign to each individual (each occupied site) on the lattice, a unique number that represents the individual's *cluster number*. This number is appended to the individual's attribute list using

```
k = 0;
latCluster = lat /. {a_, memes_List} :> {a, memes, ++k}
```

We now apply the following transformation to the lattice: each individual's cluster number is changed to equal the maximum cluster number among those of his five nearest neighbors (including himself) who have the same meme list as he has. This can be done with the following rewrite rules.

```
renumber[0, __] := 0
renumber[{a_, memes_List, c_}, res__] :=
        {a, memes, Max[c, Cases[{res}, {_, memes, d_} :> d]]}
```

The renumber rules are applied to the lattice using

```
vonNeumann[renumber, latCluster]
```

where

```
vonNeumann[func_, lat_] :=
  MapThread[func, Map[RotateRight[lat, #]&,
          {{0, 0}, {1, 0}, {0, -1}, {-1, 0}, {0, 1}}], 2]
```

If we perform this relabeling of cluster numbers repeatedly until the lattice configuration no longer changes, each individual in a cluster of individuals with the same meme list will then have the same cluster number as the highest numbered individual in that cluster. This can be done by employing the built-in function FixedPoint.

```
FixedPoint[vonNeumann[renumber, #]&, latCluster]
```

Note: The clusters are not numbered consecutively (1, 2, 3, …) by this procedure, although it is very simple to create a set of transformation rules to produce consecutive cluster numbers if that is desirable.

Combining these code fragments, we can create the following cluster identification function.

```
clusterNumbering[lattice_] :=
Module[{k, latCluster, renumber, vonNeumann},
  k = 0;
  latCluster = lattice /. {a_, memes_List} :> {a, memes, ++k};

  renumber[0, __] := 0;
  renumber[{a_, memes_List, c_}, res__] :=
      {a, memes, Max[c, Cases[{res}, {_, memes, d_} :> d]]};

  vonNeumann[func_, lat_] :=
    MapThread[func, Map[RotateRight[lat, #]&,
          {{0, 0}, {1, 0}, {0, -1}, {-1, 0}, {0, 1}}], 2];

  FixedPoint[vonNeumann[renumber, #]&, latCluster]]
```

4. How does the population density affect the results you obtained in the previous project?

5. Another variation of the cultural transmission model would incorporate the spread of a cultural value via a mass medium, such as television, so that each individual in the system, regardless of his location, either adopts a new value immediately upon being exposed to it, or adopts it with a probabilty proportional to the number of times he has been exposed to it. Implement both of these in cultural transmission models and compare the results you obtain from running both simulations.

2.6 References

Axelrod, Robert. 1997. *The Complexity of Cooperation: Agent-Based Models of Competition and Cooperation.* Princeton, NJ: Princeton University Press.

Epstein, Joshua M. and Robert Axtell. 1996. *Growing Artificial Societies.* Cambridge, MA: Brookings Institute Press/MIT Press.

2.7 Programs in the Chapter

2.7.1 cultureSpreadingShared

```
cultureSpreadingShared[n_, p_, s_, m_, t_] :=
Module[{society, RND, decide, exchange,
                                   walk, vonNeumann, GN},
  RND:= Random[Integer, {1, 4}];
  society = Table[Floor[p + Random[]], {n}, {n}] /.
       1 :→ {RND, Table[Random[Integer, {1, m}], {s}]};
  decide[{3, y_}, _, _, {1, y_}, _] := {3, y};
  decide[{2, y_}, _, {4, y_}, _, _] := {2, y};
  decide[{3, y_}, _, _, {1, z_}, _] := Module[
    {diffs = Flatten[Position[Abs[y - z], _?Positive]]},
    {3, y, Random[Integer, {y[[#]], z[[#]]}], #}&[
        diffs[[Random[Integer, {1, Length[diffs]}]]]]] /;
            Random[Integer, {1,s}] > Length[diffs]];
  decide[{2, y_}, _, {4, z_}, _, _] := Module[
    {diffs = Flatten[Position[Abs[y - z], _?Positive]]},
    {2, y, Random[Integer, {y[[#]], z[[#]]}], #}&[
        diffs[[Random[Integer, {1, Length[diffs]}]]]]] /;
            Random[Integer, {1,s}] > Length[diffs]];
  decide[a_, _, _, _, _] := a;
  exchange[{3, y_, c_, d_}, _, _, _, _] :=
                              {3, ReplacePart[y, c, d]};
  exchange[{2, y_, c_, d_}, _, _, _, _] :=
                              {2, ReplacePart[y, c, d]};
  exchange[{1, z_}, {3, y_, c_, d_}, _, _, _] :=
                              {1, ReplacePart[z, c, d]};
  exchange[{4, z_}, _, _, _, {2, y_, c_, d_}] :=
                              {4, ReplacePart[z, c, d]};
  exchange[a_, _, _, _, _] := a;
  walk[{1,a___},0,_,_,_,{4,___},_,_,_,_,_,_,_] := {RND,a};
  walk[{1,a___},0,_,_,_,_,_,_,{2,___},_,_,_,_] := {RND,a};
  walk[{1,a___},0,_,_,_,_,_,_,_,{3,___},_,_,_] := {RND,a};
  walk[{1,a___},0,_,_,_,_,_,_,_,_,_,_,_] := 0;
  walk[{2,a___},_,0,_,_,{3,___},_,_,_,_,_,_,_] := {RND,a};
  walk[{2,a___},_,0,_,_,_,{1,___},_,_,_,_,_,_] := {RND,a};
  walk[{2,a___},_,0,_,_,_,_,_,_,{4,___},_,_,_] := {RND,a};
  walk[{2,a___},_,0,_,_,_,_,_,_,_,_,_,_] := 0;
  walk[{3,a___},_,_,0,_,_,{4,___},_,_,_,_,_,_] := {RND,a};
  walk[{3,a___},_,_,0,_,_,_,{2,___},_,_,_,_,_] := {RND,a};
  walk[{3,a___},_,_,0,_,_,_,_,_,_,{1,___},_] := {RND,a};
  walk[{3,a___},_,_,0,_,_,_,_,_,_,_,_,_] := 0;
  walk[{4,a___},_,_,_,0,_,_,{1,___},_,_,_,_,_] := {RND,a};
  walk[{4,a___},_,_,_,0,_,_,{3,___},_,_,_,_,_] := {RND,a};
  walk[{4,a___},_,_,_,0,_,_,_,_,_,_,_,{2,___}] := {RND,a};
  walk[{4,a___},_,_,_,0,_,_,_,_,_,_,_,_] := 0;
  walk[{_,a___},_,_,_,_,_,_,_,_,_,_,_,_] := {RND,a};
  walk[0,{3,___},{4,___},_,_,_,_,_,_,_,_,_,_] := 0;
  walk[0,{3,___},_,{1,___},_,_,_,_,_,_,_,_,_] := 0;
  walk[0,{3,___},_,_,{2,___},_,_,_,_,_,_,_,_] := 0;
```

```
walk[0,_,{4,___},{1,___},_,_,_,_,_,_,_,_,_] := 0;
walk[0,_,{4,___},_,{2,___},_,_,_,_,_,_,_,_] := 0;
walk[0,_,_,{1,___},{2,___},_,_,_,_,_,_,_,_] := 0;
walk[0,{3,a___},_,_,_,_,_,_,_,_,_,_,_] := {RND,a};
walk[0,_,{4,a___},_,_,_,_,_,_,_,_,_,_] := {RND,a};
walk[0,_,_,{1,a___},_,_,_,_,_,_,_,_,_] := {RND,a};
walk[0,_,_,_,{2,a___},_,_,_,_,_,_,_,_] := {RND,a};
walk[0,_,_,_,_,_,_,_,_,_,_,_,_] := 0;
vonNeumann[func_, lat_] :=
  MapThread[func, Map[RotateRight[lat, #]&,
       {{0, 0}, {1, 0}, {0, -1}, {-1, 0}, {0, 1}}], 2];
GN[func_, lat_] :=
  MapThread[func, Map[RotateRight[lat, #]&,
          {{0, 0}, {1, 0}, {0, -1}, {-1, 0}, {0, 1},
           {1, -1}, {-1, -1}, {-1, 1}, {1, 1}, {2, 0},
           {0, -2}, {-2, 0}, {0, 2}}], 2];
NestList[GN[walk, vonNeumann[exchange,
                   vonNeumann[decide, #]]]&, society, t]]
```

2.7.2 cultureSpreadingCloseness

```
cultureSpreadingCloseness[n_, p_, s_, m_, t_] :=
Module[{society, RND, decide, exchange,
                              walk, vonNeumann, GN},
  RND:= Random[Integer, {1, 4}];
  society = Table[Floor[p + Random[]], {n}, {n}] /.
         1 :> {RND, Table[Random[Integer, {1, m}], {s}]};
  decide[{3, y_}, _, _, {1, y_}, _] := {3, y};
  decide[{2, y_}, _, {4, y_}, _, _] := {2, y};
  decide[{2, y_}, _, {4, z_}, _, _] :=
    Module[{r = Random[Integer, {1, s}]},
         {3, y, Random[Integer, {y[[r]], z[[r]]}], r} /;
         Random[Integer,{1, m}] > Abs[y[[r]] - z[[r]]]];
  decide[{3, y_}, _, _, {1, z_}, _] :=
    Module[{r = Random[Integer, {1, s}]},
         {3, y, Random[Integer, {y[[r]], z[[r]]}], r} /;
         Random[Integer,{1, m}] > Abs[y[[r]] - z[[r]]]];
  decide[a_, _, _, _, _] := a;
  exchange[{3, y_, c_, d_}, _, _, _, _] :=
                     {3, ReplacePart[y, c, d]};
  exchange[{2, y_, c_, d_}, _, _, _, _] :=
                     {2, ReplacePart[y, c, d]};
  exchange[{1, z_}, {3, y_, c_, d_}, _, _, _] :=
                     {1, ReplacePart[z, c, d]};
  exchange[{4, z_}, _, _, _, {2, y_, c_, d_}] :=
                     {4, ReplacePart[z, c, d]};
  exchange[a_, _, _, _, _] := a;
  walk[{1,a___},0,_,_,_,{4,___},_,_,_,_,_,_,_] := {RND,a};
  walk[{1,a___},0,_,_,_,_,_,{2,___},_,_,_,_] := {RND,a};
  walk[{1,a___},0,_,_,_,_,_,_,{3,___},_,_,_] := {RND,a};
  walk[{1,a___},0,_,_,_,_,_,_,_,_,_,_,_] := 0;
  walk[{2,a___},_,0,_,_,{3,___},_,_,_,_,_,_] := {RND,a};
  walk[{2,a___},_,0,_,_,_,{1,___},_,_,_,_,_] := {RND,a};
  walk[{2,a___},_,0,_,_,_,_,_,{4,___},_,_] := {RND,a};
```

```
        walk[{2,a___},_,0,_,_,_,_,_,_,_,_,_,_,_] := 0;
        walk[{3,a___},_,_,0,_,_,{4,___},_,_,_,_,_,_] := {RND,a};
        walk[{3,a___},_,_,0,_,_,_,{2,___},_,_,_,_,_] := {RND,a};
        walk[{3,a___},_,_,0,_,_,_,_,_,_,{1,___},_] := {RND,a};
        walk[{3,a___},_,_,0,_,_,_,_,_,_,_,_] := 0;
        walk[{4,a___},_,_,_,0,_,_,{1,___},_,_,_,_,_] := {RND,a};
        walk[{4,a___},_,_,_,0,_,_,_,{3,___},_,_,_,_] := {RND,a};
        walk[{4,a___},_,_,_,0,_,_,_,_,_,_,{2,___}] := {RND,a};
        walk[{4,a___},_,_,_,0,_,_,_,_,_,_,_] := 0;
        walk[{_,a___},_,_,_,_,_,_,_,_,_,_,_,_] := {RND,a};
        walk[0,{3,___},{4,___},_,_,_,_,_,_,_,_,_,_] := 0;
        walk[0,{3,___},_,{1,___},_,_,_,_,_,_,_,_,_] := 0;
        walk[0,{3,___},_,_,{2,___},_,_,_,_,_,_,_,_] := 0;
        walk[0,_,{4,___},{1,___},_,_,_,_,_,_,_,_,_] := 0;
        walk[0,_,{4,___},_,{2,___},_,_,_,_,_,_,_,_] := 0;
        walk[0,_,_,{1,___},{2,___},_,_,_,_,_,_,_,_] := 0;
        walk[0,{3,a___},_,_,_,_,_,_,_,_,_,_,_] := {RND,a};
        walk[0,_,{4,a___},_,_,_,_,_,_,_,_,_,_] := {RND,a};
        walk[0,_,_,{1,a___},_,_,_,_,_,_,_,_,_] := {RND,a};
        walk[0,_,_,_,{2,a___},_,_,_,_,_,_,_,_] := {RND,a};
        walk[0,_,_,_,_,_,_,_,_,_,_,_,_] := 0;
    vonNeumann[func_, lat_] :=
        MapThread[func, Map[RotateRight[lat, #]&,
            {{0, 0}, {1, 0}, {0, -1}, {-1, 0}, {0, 1}}], 2];
    GN[func_, lat_] :=
        MapThread[func, Map[RotateRight[lat, #]&,
            {{0, 0}, {1, 0}, {0, -1}, {-1, 0}, {0, 1},
             {1, -1}, {-1, -1}, {-1, 1}, {1, 1}, {2, 0},
             {0, -2}, {-2, 0}, {0, 2}}], 2];
    NestList[GN[walk, vonNeumann[exchange,
                    vonNeumann[decide, #]]]&, society, t]]
```

2.7.3 charlesBarkley

```
    charlesBarkley[n_, p_, s_, m_, t_] :=
    Module[{society, RND, exchange, walk, vonNeumann, GN},
        RND := Random[Integer, {1, 4}];
        society = Table[Floor[p + Random[]], {n}, {n}] /. 1 :>
            {RND, Random[], Table[Random[Integer, {1, m}], {s}]};
        exchange[{1, a_, u_}, {3, b_, v_}, _, _, _] :=
            {1, a, ReplacePart[u, v[[#]], #]&[
                            Random[Integer, {1, s}]]} /; a < b;
        exchange[{2, a_, u_}, _, {4, b_, v_}, _, _] :=
            {2, a, ReplacePart[u, v[[#]], #]&[
                            Random[Integer, {1, s}]]} /; a < b;
        exchange[{3, a_, u_}, _, _, {1, b_, v_}, _] :=
            {3, a, ReplacePart[u, v[[#]], #]&[
                            Random[Integer, {1, s}]]} /; a < b;
        exchange[{4, a_, u_}, _, _, _, {2, b_, v_}] :=
            {4, a, ReplacePart[u, v[[#]], #]&[
                            Random[Integer, {1, s}]]} /; a < b;
        exchange[z_, _, _, _, _] := z;
        walk[{1,a___},0,_,_,_,{4,___},_,_,_,_,_,_] := {RND,a};
        walk[{1,a___},0,_,_,_,_,{2,___},_,_,_,_,_] := {RND,a};
```

```
walk[{1,a___},0,_,_,_,_,_,_,_,{3,___},_,_,_] := {RND,a};
walk[{1,a___},0,_,_,_,_,_,_,_,_,_,_,_] := 0;
walk[{2,a___},_,0,_,_,{3,___},_,_,_,_,_,_] := {RND,a};
walk[{2,a___},_,0,_,_,_,{1,___},_,_,_,_,_] := {RND,a};
walk[{2,a___},_,0,_,_,_,_,_,_,{4,___},_,_] := {RND,a};
walk[{2,a___},_,0,_,_,_,_,_,_,_,_,_] := 0;
walk[{3,a___},_,_,0,_,_,{4,___},_,_,_,_,_] := {RND,a};
walk[{3,a___},_,_,0,_,_,_,{2,___},_,_,_,_] := {RND,a};
walk[{3,a___},_,_,0,_,_,_,_,_,_,{1,___},_] := {RND,a};
walk[{3,a___},_,_,0,_,_,_,_,_,_,_,_] := 0;
walk[{4,a___},_,_,_,0,_,_,{1,___},_,_,_,_] := {RND,a};
walk[{4,a___},_,_,_,0,_,_,_,{3,___},_,_,_] := {RND,a};
walk[{4,a___},_,_,_,0,_,_,_,_,_,_,{2,___}] := {RND,a};
walk[{4,a___},_,_,_,0,_,_,_,_,_,_,_] := 0;
walk[{_,a___},_,_,_,_,_,_,_,_,_,_,_] := {RND,a};
walk[0,{3,___},{4,___},_,_,_,_,_,_,_,_,_] := 0;
walk[0,{3,___},_,{1,___},_,_,_,_,_,_,_,_] := 0;
walk[0,{3,___},_,_,{2,___},_,_,_,_,_,_,_] := 0;
walk[0,_,{4,___},{1,___},_,_,_,_,_,_,_,_] := 0;
walk[0,_,{4,___},_,{2,___},_,_,_,_,_,_,_] := 0;
walk[0,_,_,{1,___},{2,___},_,_,_,_,_,_,_] := 0;
walk[0,{3,a___},_,_,_,_,_,_,_,_,_,_] := {RND,a};
walk[0,_,{4,a___},_,_,_,_,_,_,_,_,_] := {RND,a};
walk[0,_,_,{1,a___},_,_,_,_,_,_,_,_] := {RND,a};
walk[0,_,_,_,{2,a___},_,_,_,_,_,_,_] := {RND,a};
walk[0,_,_,_,_,_,_,_,_,_,_,_] := 0;
vonNeumann[func_, lat_] :=
  MapThread[func, Map[RotateRight[lat, #]&,
        {{0, 0}, {1, 0}, {0, -1}, {-1, 0}, {0, 1}}], 2];
GN[func_, lat_] :=
  MapThread[func, Map[RotateRight[lat, #]&,
          {{0, 0}, {1, 0}, {0, -1}, {-1, 0}, {0, 1},
           {1, -1}, {-1, -1}, {-1, 1}, {1, 1}, {2, 0},
           {0, -2}, {-2, 0}, {0, 2}}], 2];
NestList[GN[walk, vonNeumann[exchange, #]]&, society, t]]
```

Socioeconomic Transactions

"When I'm good, I'm very good. But when I'm bad, I'm better."

—Mae West in *I'm No Angel* (1933)

3

Deciding Whether to Interact

"You've got a nasty reputation, Mr. Gittes. I like that."

—John Huston in *Chinatown* (1974)

3.1 Introduction

Why are people generally honest in their dealings with others, even in the absense of a central authority to enforce good behavior? In this chapter, we model the role of ostracism as a tool for discouraging bad behavior, and thereby encouraging good behavior. We first look at two cases in which people have the ability to remember or learn about other people's bad behavior. In the first situation, people remember every individual who has *done them wrong* in a previous encounter and they refuse to interact with such a person again. In the second situation, good guys use word-of-mouth or gossip in addition to personal experience to learn who are the bad guys, and they avoid interacting with a person with a *bad rep* even once. In a final case, people will be able to *pick up on* the signals of intent (to be good or bad) emitted by each person and use those *vibes* to decide whether to interact with a stranger about whom they have no information.

3.1.1 The Prisoner's Dilemma

The most widely used theoretical model in the academic study of person-to-person interactions is known as the prisoner's dilemma (PD). In this game-theoretic model, individuals interact with one another and as a result receive payoffs. This payoff, which can take all sorts of forms (monetary, psychological, etc.), is called a *benefit* if it is positive and a *cost* if it is negative.

The classical version of the PD has two people who *must interact* and who do so *without having information about each other*. The payoffs to each person when they have the choice of being good or bad are shown in the following matrix.

		Good		Bad	
			R		*T*
Good	*R*		*S*		
			S		*P*
Bad	*T*		*P*		

According to this payoff matrix, there are four possible outcomes of the interaction:

- a win–win situation where both people are good guys and each receives a payoff of *R* (reward);
- a win–lose situation and a lose–win situation where a good guy and a bad guy interact, and the good guy receives a payoff of *S* (sucker's payoff) and the bad guy receives a payoff of *T* (temptation);
- a lose–lose situation where two bad guys interact and each receives a payoff of *P* (penalty).

It has been shown analytically that for a single interaction between two people, it is always in each person's self-interest to be uncooperative if $T > R > P > S$.

In real life, relatively few person-to-person interactions satisfy the criteria of the classic PD. In most cases, interactions are not mandatory and information about a person can be used to decide whether to interact with that person. Various sorts of information can be used for this purpose:

- a person's track record of previous behavior, learned either through personal experience or gossip;
- a person's formal and informal organizational affiliations (religious, political, charitable) serve as references; and
- a person's character traits, including changeable attributes such as shared values, appearance, or social status and unchangeable attributes, such as gender, race, or social status.

Note: Social status can be changeable in a society where wealth literally can buy social status or unchangeable in a society where genealogy determines social status.

3.2 Fool Me Once, Shame on You; Fool Me Twice, Shame on Me

In this model, an individual decides whether to interact with another person, based on knowledge about that person which has been gained through a previous first-hand interaction with the person.

3.2.1 The System

Our model uses a square n by n lattice with wraparound boundary conditions. There is a population density p of individuals occupying lattice sites and the remaining sites are empty. A fraction g of the individuals are good guys and the rest of the people are bad guys. The system evolves over a given number of time steps t.

3.2.2 Populating Society

The value of an empty site is 0.

The value of a site occupied by an individual is a five-tuple, $\{a, b, c, d, e\}$, where the list elements are:

a — an integer value between 1 and 4, indicating the direction (N, E, S, and W, respectively) faced by the individual;

b — a number (1, 2, ...) representing the individual's name;

c — 0 or 1, indicating that the individual is a bad guy (value 0) or a good guy (value 1);

d — a list of bad guys who have interacted with the individual in previous encounters.;

e — an integer value indicating the individual's resource level (the sum of the payoffs from previous time steps).

The initial system configuration consists of individuals randomly distributed on the lattice and facing randomly chosen directions. The individuals have names, zero resources, and no knowledge of the past behavior of other individuals. This configuration is created as follows.

```
k = 0;
RND := Random[Integer, {1, 4}]
society = Table[Floor[p + Random[]], {n}, {n}] /.
            1 :> {RND, ++k, Floor[g + Random[]], {}, 0}
```

3.2.3 Executing a Time Step

The following processes occur consecutively during each time step.

- Each individual who is face-to-face with another individual on an adjacent site checks his own *memory* to see if the other individual behaved badly in a previous interaction and also checks the memory of the other person to see if he is on that person's bad guy list.

Note: Allowing an individual to check if he is on the other person's memory list is more efficient computationally than having each person keep a list of all of his previous interactions.

In either case, the individual refuses to interact; otherwise, the individual interacts with the other person and if the other person behaves badly, his name is added to the individual's bad guy list. In any cases, the individual receives a payoff. All other individuals remain unchanged.

- Each individual moves to the nearest neighbor site he is facing, subject to the excluded volume constraint.

Note: Since individuals that face each other during a given time step are unable to move during that time step, the two half-steps above, as well as the partial steps in the other programs in this chapter, could be combined into a single set of rules. However, they are kept separate here in order to make the explanation of the rules more transparent.

Interacting

The rules for interaction are as follows.

The first 4 rules apply to an individual who refuses to interact with a facing individual because in a previous interaction, one or the other or both of the individuals behaved badly. The cost of opting out of an interaction is denoted *W*.

```
interact[{1, a1_, x1_, b1_, r_},
         {3, a3_, x3_, b3_, _}, _, _, _] :=
              {1, a1, x1, b1, r + W} /;
                     MemberQ[b1, a3] || MemberQ[b3, a1]
interact[{2, a2_, x2_, b2_, r_}, _,
         {4, a4_, x4_, b4_, _}, _, _] :=
              {2, a2, x2, b2, r + W} /;
                     MemberQ[b2, a4] || MemberQ[b4, a2]
interact[{3, a3_, x3_, b3_, r_}, _, _,
         {1, a1_, x1_, b1_, _}, _] :=
              {3, a3, x3, b3, r + W} /;
                     MemberQ[b3, a1] || MemberQ[b1, a3]
interact[{4, a4_, x4_, b4_, r_}, _, _, _,
         {2, a2_, x2_, b2_, _}] :=
```

```
                   {4, a4, x4, b4, r + W} /;
                      MemberQ[b4, a2] || MemberQ[b2, a4]
```

The next 16 rules apply to an individual who does interact with a facing individual. There is a rule for each of the four possible combinations of good and bad behavior, for each of the four directions the individual may face.

```
interact[{1, a1_, 1, b1_, r_},
         {3, a3_, 1, _, _}, _, _, _] :=
                               {1, a1, 1, b1, r + R}
interact[{1, a1_, 0, b1_, r_},
         {3, a3_, 1, _, _}, _, _, _] :=
                               {1, a1, 0, b1, r + T}
interact[{1, a1_, 1, b1_, r_},
         {3, a3_, 0, _, _}, _, _, _] :=
                               {1, a1, 1, Union[b1, {a3}], r + S}
interact[{1, a1_, 0, b1_, r_},
         {3, a3_, 0, _, _}, _, _, _] :=
                               {1, a1, 0, Union[b1, {a3}], r + P}
interact[{2, a2_, 1, b2_, r_}, _,
         {4, a4_, 1, _, _}, _, _] :=
                               {1, a2, 1, b2, r + R}
interact[{2, a2_, 0, b2_, r_}, _,
         {4, a4_, 1, _, _}, _, _] :=
                               {1, a2, 0, b2, r + T}
interact[{2, a2_, 1, b2_, r_}, _,
         {4, a4_, 0, _, _}, _, _] :=
                               {1, a2, 1, Union[b2, {a4}], r + S}
interact[{2, a2_, 0, b2_, r_}, _,
         {4, a4_, 0, _, _}, _, _] :=
                               {1, a2, 0, Union[b2, {a4}], r + P}
interact[{3, a3_, 1, b3_, r_}, _, _,
         {1, a1_, 1, _, _}, _] :=
                               {1, a3, 1, b3, r + R}
interact[{3, a3_, 0, b3_, r_}, _, _,
         {1, a1_, 1, _, _}, _] :=
                               {1, a3, 0, b3, r + T}
interact[{3, a3_, 1, b3_, r_}, _, _,
         {1, a1_, 0, _, _}, _] :=
                               {1, a3, 1, Union[b3, {a1}], r + S}
interact[{3, a3_, 0, b3_, r_}, _, _,
         {1, a1_, 0, _, _}, _] :=
                               {1, a3, 0, Union[b3, {a1}], r + P}
interact[{4, a4_, 1, b4_, r_}, _, _, _,
         {2, a2_, 1, _, _}] :=
                               {1, a4, 1, b4, r + R}
interact[{4, a4_, 0, b4_, r_}, _, _, _,
         {2, a2_, 1, _, _}] :=
                               {1, a4, 0, b4, r + T}
interact[{4, a4_, 1, b4_, r_}, _, _, _,
         {2, a2_, 0, _, _}] :=
                               {1, a4, 1, Union[b4, {a2}], r + S}
interact[{4, a4_, 0, b4_, r_}, _, _, _,
         {2, a2_, 0, _, _}] :=
                               {1, a4, 0, Union[b4, {a2}], r + P}
```

Each subset of 4 rules in the preceding 16 can be written more compactly as a single rule that, in turn, invokes a rule governing the payoff. The set of 16 rules can therefore be replaced by the following set of 4 rules.

```
interact[{1, a1_, x1_, b1_, r_},
         {3, a3_, x3_, _, _}, _, _, _] :=
                {1, a1, x1, Union[b1, Drop[{a3}, x3]],
                                 r + payoff[x1, x3]}
interact[{2, a2_, x2_ ,b2_, r_}, _,
         {4, a4_, x4_, _, _}, _, _] :=
                {2, a2, x2, Union[b2, Drop[{a4}, x4]],
                                 r + payoff[x2, x4]}
interact[{3, a3_, x3_, b3_, r_}, _, _,
         {1, a1_, x1_, _, _}, _] :=
                {3, a3, x3, Union[b3, Drop[{a1}, x1]],
                                 r + payoff[x3, x1]}
interact[{4, a4_, x4_, b4_, r_}, _, _, _,
         {2, a2_, x2_, _, _}] :=
                {4, a4, x4, Union[b4, Drop[{a2}, x2]],
                                 r + payoff[x4, x2]}
```

where the payoff rules are given by

```
payoff[1, 1] = R
payoff[0, 1] = T
payoff[1, 0] = S
payoff[0, 0] = P
```

There is also an interact rule which leaves empty sites and individuals who are not face-to-face with another person unchanged.

```
interact[z_, _, _, _, _] := z
```

The interact rules are applied to the lattice using the anonymous function

```
vonNeumann[interact, #]&
```

where

```
vonNeumann[func_, lat_] :=
  MapThread[func, Map[RotateRight[lat, #]&,
            {{0, 0}, {1, 0}, {0, -1}, {-1, 0}, {0, 1}}], 2]
```

Moving

The 28 rules for the movement of the individuals are given by the set of walk rules in Chapter 1.

The walk rules are applied to the lattice sites using the anonymous function

```
GN[walk, vonNeumann[interact, #]]&
```

where

```
GN[func_, lat_] :=
  MapThread[func, Map[RotateRight[lat, #]&,
            {{0, 0}, {1, 0}, {0, -1}, {-1, 0}, {0, 1},
             {1, -1}, {-1, -1}, {-1, 1}, {1, 1}, {2, 0},
             {0, -2}, {-2, 0}, {0, 2}}], 2]
```

3.2.4 Evolving the System

The system evolves over *t* time steps, starting with the initial lattice configuration, society, using the following nesting operation.

```
NestList[vonNeumann[interact, GN[walk, #]]&, society, t]
```

3.2.5 The Program

```
onceBurntTwiceShy[n_, p_, g_,
                      {P_, R_, S_, T_, W_}, t_] :=
Module[{RND, k = 0, society, interact, payoff,
                             walk, vonNeumann, GN},

  RND := Random[Integer, {1, 4}];
  society = Table[Floor[p + Random[]], {n}, {n}] /.
            1 :> {RND, ++k, Floor[g + Random[]], {}, 0};

  payoff[1, 1] = R;
  payoff[0, 1] = T;
  payoff[1, 0] = S;
  payoff[0, 0] = P;

  interact[{1, a1_, x1_, b1_, r_},
           {3, a3_, x3_, b3_, _}, _, _, _] :=
             {1, a1, x1, b1, r + W} /;
                 MemberQ[b1, a3] || MemberQ[b3, a1];
  interact[{2, a2_, x2_, b2_, r_}, _,
           {4, a4_, x4_, b4_, _}, _, _] :=
             {2, a2, x2, b2, r + W} /;
                 MemberQ[b2, a4] || MemberQ[b4, a2];
  interact[{3, a3_, x3_, b3_, r_}, _, _,
           {1, a1_, x1_, b1_, _}, _] :=
             {3, a3, x3, b3, r + W} /;
                 MemberQ[b3, a1] || MemberQ[b1, a3];
  interact[{4, a4_, x4_, b4_, r_}, _, _, _,
           {2, a2_, x2_, b2_, _}] :=
             {4, a4, x4, b4, r + W} /;
                 MemberQ[b4, a2] || MemberQ[b2, a4];
  interact[{1, a1_, x1_, b1_, r_},
           {3, a3_, x3_, _, _}, _, _, _] :=
             {1, a1, x1, Union[b1, Drop[{a3}, x3]],
                             r + payoff[x1, x3]};
```

```
interact[{2, a2_, x2_ ,b2_, r_}, _,
         {4, a4_, x4_, _, _}, _, _] :=
              {2, a2, x2, Union[b2, Drop[{a4}, x4]],
                                  r + payoff[x2, x4]};
interact[{3, a3_, x3_, b3_, r_}, _, _,
         {1, a1_, x1_, _, _}, _] :=
              {3, a3, x3, Union[b3, Drop[{a1}, x1]],
                                  r + payoff[x3, x1]};
interact[{4, a4_, x4_, b4_, r_}, _, _, _,
         {2, a2_, x2_, _, _}] :=
              {4, a4, x4, Union[b4, Drop[{a2}, x2]],
                                  r + payoff[x4, x2]};
interact[z_, _, _, _, _] := z;

(* walk rules go here *)

vonNeumann[func_, lat_] :=
  MapThread[func, Map[RotateRight[lat, #]&,
       {{0, 0}, {1, 0}, {0, -1}, {-1, 0}, {0, 1}}], 2];
GN[func_, lat_] :=
  MapThread[func, Map[RotateRight[lat, #]&,
          {{0, 0}, {1, 0}, {0, -1}, {-1, 0}, {0, 1},
           {1, -1}, {-1, -1}, {-1, 1}, {1, 1}, {2, 0},
           {0, -2}, {-2, 0}, {0, 2}}], 2];

NestList[vonNeumann[interact, GN[walk, #]]&, society, t]]
```

3.2.6 Running the Simulation

We first run the onceBurntTwiceShy program on a 50 by 50 lattice having a 70% population density equally divided between good guys and bad guys and the payoff values used by Axelrod of $\{1, 3, 0, 5, 0\}$ over 500 time steps.

```
SeedRandom[21]
results = onceBurntTwiceShy[50, 0.7, 0.5,
                              {1, 3, 0, 5, 0}, 500];
```

For the lattice configuration at a particular time step, the avgResourceLevel function computes the average of all the resource levels of individuals following a given strategy.

```
avgResourceLevel[lat_, strat_] :=
  (Apply[Plus, #]/Length[#])&[
              Cases[lat, {_, _, strat, _, x_} :> x, 2]];
```

Using this function with the results of the program run, the average resource levels for each type of individual over time is shown in the following.

```
MultipleListPlot[
  Map[avgResourceLevel[#, 1]&, results],
  Map[avgResourceLevel[#, 0]&, results],
  SymbolShape -> None,
  PlotJoined -> True,
```

```
PlotLegend → {"Good guy", "Bad guy"},
AxesLabel → {"Time", "Resource level"},
LegendShadow → {-0.05, -0.05}];
```

Resource level

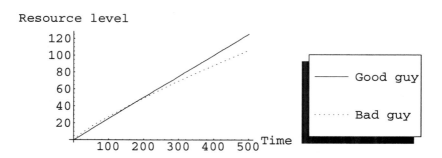

It appears that being able to remember past encounters with bad guys enables *good to triumph over evil* in terms of resource level but only in the long run. This is because the knowledge of bad behavior in the model is kept private—no one tells anyone else about it. We next see whether using public knowledge or gossip in addition to private knowledge can *turn the tables* on the bad guys even sooner.

3.3 I Heard It Through the Grapevine

We can allow individuals who interact with one another to share information or *gossip* about the bad behavior of other individuals. This exchange results in individuals being excluded or *ostracized* from interactions with people with whom they have never interacted in the past and so imposes an even higher penalty on past bad behavior than the previous model.

Interacting

The onceBurntTwiceShy program can be very easily modified to include gossiping by

- redefining the fourth component of the list characterizing an individual (i.e., characterizing a site that is occupied by an individual) to be a list of the bad guys who have interacted previously with either the individual or with one of the good guys with whom the individual has interacted, and

- making a simple change in the four interact rules for individuals who do interact (the interact rules for individuals who don't interact remain the same).

```
interact[{1, a1_, x1_, b1_, r_},
         {3, a3_, x3_, b3_, _}, _, _, _] :=
             {1, a1, x1, Union[b1, b3, Drop[{a3}, x3]],
                                r + payoff[x1, x3]}
```

```
interact[{2, a2_, x2_ ,b2_, r_}, _,
        {4, a4_, x4_, b4_, _}, _, _] :=
               {2, a2, x2, Union[b2, b4, Drop[{a4}, x4]],
                                     r + payoff[x2, x4]}
interact[{3, a3_, x3_, b3_, r_}, _, _,
        {1, a1_, x1_, b1_, _}, _] :=
               {3, a3, x3, Union[b3, b1, Drop[{a1}, x1]],
                                     r + payoff[x3, x1]}
interact[{4, a4_, x4_, b4_, r_}, _, _, _,
        {2, a2_, x2_, b2_, _}] :=
               {4, a4, x4, Union[b4, b2, Drop[{a2}, x2]],
                                     r + payoff[x4, x2]}
```

3.3.1 The Program

```
badRep[n_, p_, g_, {P_, R_, S_, T_, W_}, t_] :=
Module[{RND, k = 0, society, interact, payoff,
                              walk, vonNeumann, GN},
  RND := Random[Integer, {1, 4}];
  society = Table[Floor[p + Random[]], {n}, {n}] /.
            1 :> {RND, ++k, Floor[g + Random[]], {}, 0};
  payoff[1, 1] = R;
  payoff[0, 1] = T;
  payoff[1, 0] = S;
  payoff[0, 0] = P;

  interact[{1, a1_, x1_, b1_, r_},
          {3, a3_, x3_, b3_, _}, _, _, _] :=
                 {1, a1, x1, b1, r + W} /;
                     MemberQ[b1, a3] || MemberQ[b3, a1];
  interact[{2, a2_, x2_, b2_, r_}, _,
          {4, a4_, x4_, b4_, _}, _, _] :=
                 {2, a2, x2, b2, r + W} /;
                     MemberQ[b2, a4] || MemberQ[b4, a2];
  interact[{3, a3_, x3_, b3_, r_}, _, _,
          {1, a1_, x1_, b1_, _}, _] :=
                 {3, a3, x3, b3, r + W} /;
                     MemberQ[b3, a1] || MemberQ[b1, a3];
  interact[{4, a4_, x4_, b4_, r_}, _, _, _,
          {2, a2_, x2_, b2_, _}] :=
                 {4, a4, x4, b4, r + W} /;
                     MemberQ[b4, a2] || MemberQ[b2, a4];
  interact[{1, a1_, x1_, b1_, r_},
          {3, a3_, x3_, b3_, _}, _, _, _] :=
                 {1, a1, x1, Union[b1, b3, Drop[{a3}, x3]],
                                       r + payoff[x1, x3]};
  interact[{2, a2_, x2_ ,b2_, r_}, _,
          {4, a4_, x4_, b4_, _}, _, _] :=
                 {2, a2, x2, Union[b2, b4, Drop[{a4}, x4]],
                                       r + payoff[x2, x4]};
  interact[{3, a3_, x3_, b3_, r_}, _, _,
          {1, a1_, x1_, b1_, _}, _] :=
                 {3, a3, x3, Union[b3, b1, Drop[{a1}, x1]],
                                       r + payoff[x3, x1]};
```

```
interact[{4, a4_, x4_, b4_, r_}, _, _, _,
       {2, a2_, x2_, b2_, _}] :=
            {4, a4, x4, Union[b4, b2, Drop[{a2}, x2]],
                                  r + payoff[x4, x2]};
interact[z_, _, _, _, _] := z;

(* walk rules go here *)

vonNeumann[func_, lat_] :=
  MapThread[func, Map[RotateRight[lat, #]&,
       {{0, 0}, {1, 0}, {0, -1}, {-1, 0}, {0, 1}}], 2];

GN[func_, lat_] :=
  MapThread[func, Map[RotateRight[lat, #]&,
       {{0, 0}, {1, 0}, {0, -1}, {-1, 0}, {0, 1},
        {1, -1}, {-1, -1}, {-1, 1}, {1, 1}, {2, 0},
        {0, -2}, {-2, 0}, {0, 2}}], 2];

NestList[vonNeumann[interact, GN[walk, #]]&, society, t]]
```

3.3.2 Running the Simulation

We first run the program on a 50 by 50 lattice having a 70% population density equally divided between good guys and bad guys and payoff values of {-1, 1, -2, 2, 0} over 500 time steps.

```
SeedRandom[17]
results = badRep[50, 0.7, 0.5, {-1, 1, -2, 2, 0}, 500];
```

For the lattice configuration at a particular time step, the avgResourceLevel function computes the average of all the resource levels of individuals following a given strategy, and the avgMemoryLength function calculates the average number of bad people whose bad rep is known to individuals following a given strategy.

```
avgResourceLevel[lat_, strat_] :=
  (Apply[Plus, #]/Length[#])&[
             Cases[lat, {_, _, strat, _, x_} :> x, 2]]
avgMemoryLength[lat_, strat_] :=
  (Apply[Plus, Map[Length, #]]/Length[#])&[
             Cases[lat, {_, _, strat, x_, _} :> x, 2]]
```

We can apply these two functions to the lattice configurations for each time step and use the results to create average memory length and average resource level lists for each type of individual, using

```
goodMemoryLength =
            Map[avgMemoryLength[#, 1]&, results];
goodResourceLevel =
            Map[avgResourceLevel[#, 1]&, results];
badMemoryLength =
            Map[avgMemoryLength[#, 0]&, results];
badResourceLevel =
            Map[avgResourceLevel[#, 0]&, results];
```

We can plot the resource levels of good and bad guys over time using

```
MultipleListPlot[goodResourceLevel, badResourceLevel,
    SymbolShape → None,
    PlotJoined → True,
    PlotLegend → {"Good guy", "Bad guy"},
    AxesLabel → {"Time", "Resource Level"},
    LegendShadow → {-.05, -.05}];
```

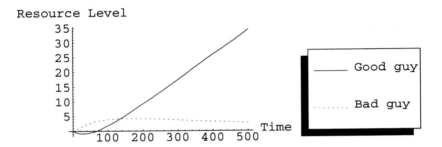

Initially, the resource levels of the bad guys increase and the resource levels of the good guys fall. However, eventually, the fortunes of these two types of people reverse their trends and the good guys' resources climb and the bad guys' resources decline.

Apparently, the bad guys initially gain at the expense of the good guys whom they are victimizing but, as time passes, the word gets around (gossip spreads) about who is a bad guy and therefore these bad guys are ostracized from interacting with good guys, to their detriment.

The memory list length values of the good and bad guys over time can be plotted using

```
MultipleListPlot[goodMemoryLength, badMemoryLength,
    SymbolShape → None,
    PlotJoined → True,
    PlotLegend → {"Good guy",  "Bad guy"},
    AxesLabel → {"Time", "Memory List Length"},
    LegendShadow → {-.05,  -.05}];
```

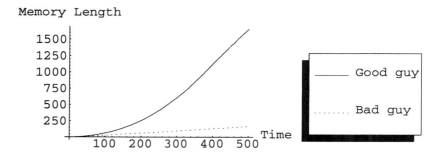

Since only the names of bad guys who interact with good guys are added to a person's list, it is not surprising that good guys learn more than bad guys about who are the bad guys.

3.4 Signals

"I suggest you deactivate your emotion chip for now."

—Patrick Stewart in *Star Trek: First Contact* (1996)

In both the onceBurntTwiceShy and badRep models, bad behavior returns an immediate benefit to a bad guy but the side effect of being placed on a good guy's list of bad guys results in a bad guy being limited to one interaction, at most, with a given individual. Hence, in these models, bad people inevitably *come to a bad end* as no mechanism is available to allow a person to escape his previous past behavior and redeem himself. It's similar to having a *scarlet A* permanently tattooed on your forehead.

These simple models of social exclusion or shunning may account for the tendency of people living in an isolated close-knit community to be cooperative and friendly with one another. However, in a society where people are mobile, the situation is different and using personal experience and reputation may be inadequate tools for discouraging bad behavior. People tend to become *lost in the crowd* in such a society and as a result they cross paths with other people about whom they know nothing, either personally or through others. In this case, the decision about whether to interact with another person must be made on some other basis.

When we consider interacting with a person about whom we can obtain no information, we must resort to *sizing up* the other person and making a best guess as to how he will behave in an interaction. It has been proposed [Frank 1988] that individuals emit various *signals* or *vibes* that allow others to anticipate the person's predisposition (nonbinding commitment) to behave in a certain way.

In this section, we focus on the use of the signals given off by another person to decide whether to interact with that person.

3.4.1 The System

Our model uses a square n by n lattice with wraparound boundary conditions. There is a population density p of individuals occupying lattice sites and the remaining sites stay empty. A fraction g of the individuals are good guys and the rest of the people are bad guys. The system evolves over a given number of time steps t.

3.4.2 Populating Society

The value of an empty site is 0.

The value of a site occupied by an individual is a four-tuple, {direction, strategy, signal, resources}, where the list elements are:

- *direction faced*, an integer value between 1 and 4, indicating the direction (N, E, S, and W, respectively) faced by the individual;

- *behavior*, 0 or 1, indicating that the individual is a bad guy (value 0) or a good guy (value 1);

- *signal strength*, a real number between 0 and 1, indicating the signal strength emitted by the individual (the higher the value, the more likely it is that there will be interaction with the individual);

- *resource level*, an integer value indicating the individual's resource level (the sum of the payoffs from previous time steps).

In the initial system configuration, individuals are randomly distributed on the lattice, facing randomly chosen directions. The individuals have zero resources and their signals have randomly chosen strengths. This configuration is created as follows.

```
RND := Random[Integer, {1, 4}]
society = Table[Floor[p + Random[]], {n}, {n}] /.
          1 :→ {RND, Floor[g + Random[]], Random[], 0}
```

3.4.3 Executing a Time Step

The following processes occur consecutively during each time step:

- Each individual who is face-to-face with another individual on an adjacent site observes the signals emanating from that individual and based on the strength of the *good vibes* he feels from the other individual, decides if he should interact or opt out (the stronger the signal, the greater the likelihood of interacting).

- Facing individuals that have agreed to interact with one another do so while other facing individuals do not interact.

- Each individual moves to the nearest neighbor site it is facing, subject to the excluded volume constraint.

Deciding

The probability that an individual who is facing another individual will decide to interact with that neighbor is proportional to the strength of the signal emitted by the facing individual. The rule set for deciding this is given by

```
decide[{1, a_, x_, b_}, {3, _, y_, _}, _, _, _] :=
                        {1, a, x, b, Floor[y + Random[]]}
decide[{2, a_, x_, b_}, _, {4, _, y_, _}, _, _] :=
                        {2, a, x, b, Floor[y + Random[]]}
decide[{3, a_, x_, b_}, _, _, {1, _, y_, _}, _] :=
                        {3, a, x, b, Floor[y + Random[]]}
decide[{4, a_, x_, b_}, _, _, _, {2, _, y_, _}] :=
                        {4, a, x, b, Floor[y + Random[]]}
decide[z_, _, _, _, _] := z
```

where the value of the last element in the five-tuple list created by the first four rules is either 1, representing a decision to interact, or 0, representing a decision to opt out.

The decide rules are applied to the lattice using the anonymous function

```
vonNeumann[decide, #]&
```

where

```
vonNeumann[func_, lat_] :=
  MapThread[func, Map[RotateRight[lat, #]&,
          {{0, 0}, {1, 0}, {0, -1}, {-1, 0}, {0, 1}}]], 2]
```

Interacting

Once the decision to interact or not has been made by each individual, facing individuals can proceed to interact or opt out, using the following rules.

```
interact[{3, a_, x_, b_, k_}, _, _,
          {1, d_, _, _, h_}, _] :=
                    {3, a, x, b + costbenefit[a, d, (k * h)]}
interact[{2, a_, x_, b_, k_}, _,
          {4, d_, _, _, h_}, _, _] :=
                    {2, a, x, b + costbenefit[a, d, (k * h)]}
interact[{1, a_, x_, b_, k_},
          {3, d_, _, _, h_}, _, _, _] :=
                    {1, a, x, b + costbenefit[a, d, (k * h)]}
interact[{4, a_, x_, b_, k_}, _, _, _,
          {2, d_, _, _, h_}] :=
                    {4, a, x, b + costbenefit[a, d, (k * h)]}
interact[z_, _, _, _, _] := z
```

where

```
costbenefit[1, 1, 1] = R
costbenefit[0, 1, 1] = T
costbenefit[1, 0, 1] = S
costbenefit[0, 0, 1] = P
costbenefit[_, _, 0] = W
```

The third argument of the costbenefit function indicates if there is an interaction. A value of 1 for this argument indicates that interaction does take place with the usual payoffs or R, T, S or P, and a value of 0 indicates that one or both of the individuals has decided to opt out, resulting in a cost of W to each of the noninteracting parties.

The interact rules are applied to the lattice using the anonymous function

```
vonNeumann[interact, vonNeumann[decide, #]]&
```

Moving

The 28 rules for the movement of the individuals are given by the set of walk rules in Chapter 1.

The walk rules are applied to the lattice sites using the anonymous function

```
GN[walk, vonNeumann[interact, vonNeumann[decide, #]]]&
```

where

```
GN[func_, lat_] :=
  MapThread[func, Map[RotateRight[lat, #]&,
            {{0, 0}, {1, 0}, {0, -1}, {-1, 0}, {0, 1},
             {1, -1}, {-1, -1}, {-1, 1}, {1, 1}, {2, 0},
             {0, -2}, {-2, 0}, {0, 2}}], 2]
```

3.4.4 Evolving the System

The system evolves over *t* time steps, starting with the initial lattice configuration, society, using the following nesting operation.

```
NestList[GN[walk, vonNeumann[interact,
                 vonNeumann[decide, #]]]&, society, t]
```

3.4.5 The Program

```
vibes[n_, p_, g_, {P_, R_, S_, T_, W_}, t_] :=
Module[{RND, society, decide, interact, costbenefit,
                              walk, vonNeumann, GN},

  RND := Random[Integer, {1, 4}];
  society = Table[Floor[p + Random[]], {n}, {n}] /.
            1 :> {RND, Floor[g + Random[]], Random[], 0};

  decide[{1, a_, x_, b_}, {3, _, y_, _}, _, _, _] :=
                {1, a, x, b, Floor[y + Random[]]};
  decide[{2, a_, x_, b_}, _, {4, _, y_, _}, _, _] :=
                {2, a, x, b, Floor[y + Random[]]};
  decide[{3, a_, x_, b_}, _, _, {1, _, y_, _}, _] :=
                {3, a, x, b, Floor[y + Random[]]};
  decide[{4, a_, x_, b_}, _, _, _, {2, _, y_, _}] :=
                {4, a, x, b, Floor[y + Random[]]};
  decide[z_, _, _, _, _] := z;

  costbenefit[1, 1, 1] = R;
  costbenefit[0, 1, 1] = T;
  costbenefit[1, 0, 1] = S;
  costbenefit[0, 0, 1] = P;
  costbenefit[_, _, 0] = W;

  interact[{3, a_, x_, b_, k_}, _, _,
        {1, d_, _, _, h_}, _] :=
              {3, a, x, b + costbenefit[a, d, (k * h)]};
  interact[{2, a_, x_, b_, k_}, _,
        {4, d_, _, _, h_}, _, _] :=
              {2, a, x, b + costbenefit[a, d, (k * h)]};
  interact[{1, a_, x_, b_, k_},
        {3, d_, _, _, h_}, _, _, _] :=
              {1, a, x, b + costbenefit[a, d, (k * h)]};
  interact[{4, a_, x_, b_, k_}, _, _, _,
        {2, d_, _, _, h_}] :=
              {4, a, x, b + costbenefit[a, d, (k * h)]};
  interact[z_, _, _, _, _] := z;

  (* walk rules go here *)
```

```
vonNeumann[func_, lat_] :=
  MapThread[func, Map[RotateRight[lat, #]&,
        {{0, 0}, {1, 0}, {0, -1}, {-1, 0}, {0, 1}}], 2];

GN[func_, lat_] :=
  MapThread[func, Map[RotateRight[lat, #]&,
          {{0, 0}, {1, 0}, {0, -1}, {-1, 0}, {0, 1},
           {1, -1}, {-1, -1}, {-1, 1}, {1, 1}, {2, 0},
           {0, -2}, {-2, 0}, {0, 2}}], 2];

NestList[GN[walk, vonNeumann[interact,
              vonNeumann[decide, #]]]&, society, t]];
```

3.4.6 Running the Simulation

We run the program for a 50 by 50 lattice having a 70% population density equally divided between good guys and bad guys and payoff values of {-1, 1, -2, 2, 0} over 500 time steps.

```
SeedRandom[29]
results = vibes[50, 0.7, 0.5, {-1, 1, -2, 2, 0}, 500];
```

The following scatter plots show the distribution of resources after the last time step of the simulation for a given type of individual. Every point on the graph indicates an agent's signal value and resource level.

```
Show[Graphics[{
  RGBColor[0, 0, 1],
  Cases[Last[results],
      {_, 1, a_, r_} :> Point[{a, r}], 2]}],
  Axes → True,
  AxesLabel → {"Signal","Resource level"},
  PlotLabel → "Good guys"];

Show[Graphics[{
  RGBColor[1, 0, 0],
  Cases[Last[results],
      {_, 0, a_, r_} :> Point[{a, r}], 2]}],
  Axes → True,
  AxesLabel → {"Signal","Resource level"},
  PlotLabel → "Bad guys"];
```

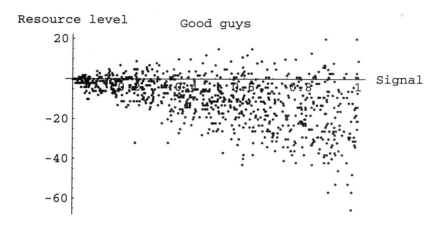

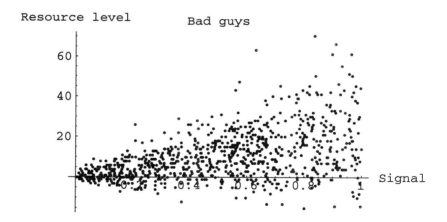

These plots appear to indicate that low signal values tend to result in small resource levels. Moreover, as an individual's signal value increases, his behavior becomes increasingly important.

3.5 Computer Simulation Projects

1. In the simple models presented in this chapter, only acts of bad behavior are used in deciding whether to interact with an individual. It is simple to incorporate acts of good behavior into the decision by having the agents keep a list of good people and to use that list, either by itself or in conjunction with the bad guy list, in deciding whether to interact. (This brings to mind the lyrics, "making a list, checking it twice, going to find out who's naughty or nice" from the song "Santa Claus Is Coming To Town.") Do this and see how it affects the results.

2. Consider a model in which individuals play with people they know to be good, don't play with people they know to be bad, and otherwise play with probability p. How does the system's behavior change for different values of p?

3. What is the effect of having a limited memory of past behavior? Create a variant of the model where an individual's bad guy list can only be a certain length, and once that length is reached, any interaction with a bad guy results in the addition of the bad guy's name to the end of the list and the removal of the first name in the list. Observe the effect on the resource levels of good guys and bad guys. As another variant, suppose that good guys have better memories (so that their bad guy list has a longer maximum length) than do bad guys. How does this compare to the case when bad guys have better memories than good guys? Yet another variant of memory list limitations to consider is to give the population a random distribution of maximum memory list lengths. Finally, suppose that each individual clears his memory (empties his bad guy list) with some probability p on each time step. How does this affect the importance of memory limitations?

4. Kitcher [Batali and Kitcher 1995] has proposed an iterated optional prisoner's dilemma game with the standard T, R, P, S, interaction payoffs and the W opt-out payoff consisting of five types of individuals:

- the solo person who never interacts;
- the undiscriminating altruist who is always a good guy and always interacts;
- the undiscriminating defector who is always a bad guy and always interacts;
- the discriminating altruist who is always a good guy and interacts with any other person who is willing to interact, provided that the person has not previously behaved badly with him—otherwise he opts out;
- the discriminating defector is always a bad guy and interacts with any other person who is willing to interact, provided that the person has not previously behaved badly with him—otherwise he opts out.

Implement this model in a program and observe its behavior over time.

5. In the vibes program, the number of decide rules can be reduced from five to three by having the determination by each of two facing individuals to interact or not be made by just one of the individuals as was done in the cultural transmission program in Chapter 2.

The computation is done as follows: if the signal strength of the south (east) facing individual is x and the signal strength of the north (west) facing individual is y, then we write a rule that updates the four-tuple value of the south (east) facing individual to a five-tuple in which an additional element has been added to the end of the list. The value of this last element is given by

```
Floor[x + Random[]] * Floor[y + Random[]]
```

which will equal 1 if both individuals have independently decided to interact and 0 if either one or both of the individuals have decided to opt out of the interaction.

The three rules for this decision-making process are written as:

```
decide[{3, a_, x_, b_}, _, _, {1, _, y_, _}, _] :=
    {3, a, x, b, Floor[x + Random[]] * Floor[y + Random[]]}
decide[{2, a_, x_, b_}, _, {4, _, y_, _}, _, _] :=
    {2, a, x, b, Floor[x + Random[]] * Floor[y + Random[]]}
decide[z_, _, _, _, _] := z
```

If these decide rules are used, then the appropriate interact rules are given by

```
interact[{3, a_, x_, b_, k_}, _, _, {1, d_, _, _}, _] :=
                        {3, a, x, b + costbenefit[a, d, k]}
interact[{2, a_, x_, b_, k_}, _, {4, d_, _, _}, _, _] :=
                        {2, a, x, b + costbenefit[a, d, k]}
interact[{1, a_, x_, b_}, {3, d_, _, _, k_}, _, _, _] :=
                        {1, a, x, b + costbenefit[a, d, k]}
interact[{4, a_, x_, b_}, _, _, _, {2, d_, _, _, k_}] :=
                        {4, a, x, b + costbenefit[a, d, k]}
interact[z_, _, _, _, _] := z
```

where

```
costbenefit[1, 1, 1] = R
costbenefit[0, 1, 1] = T
costbenefit[1, 0, 1] = S
costbenefit[0, 0, 1] = P
costbenefit[_, _, 0] = W
```

Use a variety of criteria, such as aesthetics and speed, to decide which rule set is preferable to use.

6. The usefulness of signaling depends not only on the ability of individuals to send a signal, but also on the ability of an observer to correctly interpret the signal being sent. In the vibes program, the decide rules assume that all individuals have the same signal detection ability so that each individual decides if he should interact or not based solely on the strength of the signal he receives from another individual. Modify the vibes model to endow individuals with varying abilities to detect and interpret the *first impressions* made by others.

3.6 References

Batali, John and Philip Kitcher. 1995. "Evolution of Altruism in Optional and Compulsory Games." *Journal of Theoretical Biology* 175: 161–171.

Enquist, Magnus and Olof Leimat. 1993. "The Evolution of Cooperation in Mobile Organisms." *Animal Behavior* 45: 747–757.

Frank, Robert H. 1988. *Passions Within Reason: The Strategic Role of the Emotions*. New York, NY: W.W. Norton.

Frank emphasizes the use of the visual display of emotions, such as facial expressions or body language as a signaling device that allows others to anticipate the person's predisposition to behave in a certain way. He says that emotions are especially useful for providing a fairly accurate measure of a person's intentions because emotional responses such as indignation, humor, fear, guilt, remorse, compassion, and the like (what Adam Smith called moral sentiments) are generally beyond a person's purposeful control, making them difficult to fake consistently and rendering a person's behavior automatic and consistent. As a result, emotionally based behavior is at least somewhat predictable by others, and therefore a person's emotional displays can be used by another person to decide whether to interact with him.

Thaler, Richard H. 1992. *The Winner's Curse: Paradoxes and Anomalies of Economic Life*. The Free Press, chap. 2, 6–20. Reprinted from Robyn M. Dawes and Richard H. Thaler, 1988, "Cooperation." *Journal of Economic Persepctives* 2 (Summer): 187–197.

3.7 Programs in the Chapter

3.7.1 onceBurntTwiceShy

```
onceBurntTwiceShy[n_, p_, g_,
                    {P_, R_, S_, T_, W_}, t_] :=
Module[{RND, k = 0, society, interact, payoff,
                            walk, vonNeumann, GN},
  RND := Random[Integer, {1, 4}];
  society = Table[Floor[p + Random[]], {n}, {n}] /.
            1 :> {RND, ++k, Floor[g + Random[]], {}, 0};
  payoff[1, 1] = R;
  payoff[0, 1] = T;
  payoff[1, 0] = S;
  payoff[0, 0] = P;
  interact[{1, a1_, x1_, b1_, r_},
           {3, a3_, x3_, b3_, _}, _, _, _] :=
              {1, a1, x1, b1, r + W} /;
                 MemberQ[b1, a3] || MemberQ[b3, a1];
  interact[{2, a2_, x2_, b2_, r_}, _,
           {4, a4_, x4_, b4_, _}, _, _] :=
              {2, a2, x2, b2, r + W} /;
                 MemberQ[b2, a4] || MemberQ[b4, a2];
  interact[{3, a3_, x3_, b3_, r_}, _, _,
           {1, a1_, x1_, b1_, _}, _] :=
              {3, a3, x3, b3, r + W} /;
                 MemberQ[b3, a1] || MemberQ[b1, a3];
  interact[{4, a4_, x4_, b4_, r_}, _, _, _,
           {2, a2_, x2_, b2_, _}] :=
              {4, a4, x4, b4, r + W} /;
                 MemberQ[b4, a2] || MemberQ[b2, a4];
  interact[{1, a1_, x1_, b1_, r_},
           {3, a3_, x3_, _, _}, _, _, _] :=
              {1, a1, x1, Union[b1, Drop[{a3}, x3]],
                              r + payoff[x1, x3]};
  interact[{2, a2_, x2_ ,b2_, r_}, _,
           {4, a4_, x4_, _, _}, _, _] :=
              {2, a2, x2, Union[b2, Drop[{a4}, x4]],
                              r + payoff[x2, x4]};
  interact[{3, a3_, x3_, b3_, r_}, _, _,
           {1, a1_, x1_, _, _}, _] :=
              {3, a3, x3, Union[b3, Drop[{a1}, x1]],
                              r + payoff[x3, x1]};
  interact[{4, a4_, x4_, b4_, r_}, _, _, _,
           {2, a2_, x2_, _, _}] :=
              {4, a4, x4, Union[b4, Drop[{b2}, x2]],
                              r + payoff[x4, x2]};
  interact[z_, _, _, _, _] := z;
  walk[{1,a___},0,_,_,_,{4,___},_,_,_,_,_,_,_] := {RND,a};
  walk[{1,a___},0,_,_,_,_,_,_,{2,___},_,_,_,_] := {RND,a};
  walk[{1,a___},0,_,_,_,_,_,_,_,_,{3,___},_,_,_] := {RND,a};
  walk[{1,a___},0,_,_,_,_,_,_,_,_,_,_,_,_] := 0;
```

```
walk[{2,a___},_,0,_,_,{3,___},_,_,_,_,_,_,_] := {RND,a};
walk[{2,a___},_,0,_,_,_,{1,___},_,_,_,_,_,_] := {RND,a};
walk[{2,a___},_,0,_,_,_,_,_,_,_,{4,___},_,_] := {RND,a};
walk[{2,a___},_,0,_,_,_,_,_,_,_,_,_] := 0;
walk[{3,a___},_,_,0,_,_,{4,___},_,_,_,_,_] := {RND,a};
walk[{3,a___},_,_,0,_,_,_,{2,___},_,_,_,_] := {RND,a};
walk[{3,a___},_,_,0,_,_,_,_,_,_,{1,___},_] := {RND,a};
walk[{3,a___},_,_,0,_,_,_,_,_,_,_,_] := 0;
walk[{4,a___},_,_,_,0,_,_,{1,___},_,_,_,_] := {RND,a};
walk[{4,a___},_,_,_,0,_,_,{3,___},_,_,_,_] := {RND,a};
walk[{4,a___},_,_,_,0,_,_,_,_,_,_,{2,___}] := {RND,a};
walk[{4,a___},_,_,_,0,_,_,_,_,_,_,_] := 0;
walk[{_,a___},_,_,_,_,_,_,_,_,_,_,_,_] := {RND,a};
walk[0,{3,___},{4,___},_,_,_,_,_,_,_,_,_] := 0;
walk[0,{3,___},_,{1,___},_,_,_,_,_,_,_,_] := 0;
walk[0,{3,___},_,_,{2,___},_,_,_,_,_,_,_] := 0;
walk[0,_,{4,___},{1,___},_,_,_,_,_,_,_,_] := 0;
walk[0,_,{4,___},_,{2,___},_,_,_,_,_,_,_] := 0;
walk[0,_,_,{1,___},{2,___},_,_,_,_,_,_,_] := 0;
walk[0,{3,a___},_,_,_,_,_,_,_,_,_,_,_] := {RND,a};
walk[0,_,{4,a___},_,_,_,_,_,_,_,_,_,_] := {RND,a};
walk[0,_,_,{1,a___},_,_,_,_,_,_,_,_,_] := {RND,a};
walk[0,_,_,_,{2,a___},_,_,_,_,_,_,_,_] := {RND,a};
walk[0,_,_,_,_,_,_,_,_,_,_,_] := 0;
vonNeumann[func_, lat_] :=
  MapThread[func, Map[RotateRight[lat, #]&,
        {{0, 0}, {1, 0}, {0, -1}, {-1, 0}, {0, 1}}], 2];
GN[func_, lat_] :=
  MapThread[func, Map[RotateRight[lat, #]&,
        {{0, 0}, {1, 0}, {0, -1}, {-1, 0}, {0, 1},
         {1, -1}, {-1, -1}, {-1, 1}, {1, 1}, {2, 0},
         {0, -2}, {-2, 0}, {0, 2}}], 2];
NestList[vonNeumann[interact, GN[walk, #]]&, society, t]]
```

3.7.2 badRep

```
badRep[n_, p_, g_, {P_, R_, S_, T_, W_}, t_] :=
Module[{RND, k = 0, society, interact, payoff,
                              walk, vonNeumann, GN},
  RND := Random[Integer, {1, 4}];
  society = Table[Floor[p + Random[]], {n}, {n}] /.
          1 :> {RND, ++k, Floor[g + Random[]], {}, 0};
  payoff[1, 1] = R;
  payoff[0, 1] = T;
  payoff[1, 0] = S;
  payoff[0, 0] = P;
  interact[{1, a1_, x1_, b1_, r_},
          {3, a3_, x3_, b3_, _}, _, _, _] :=
              {1, a1, x1, b1, r + W} /;
                  MemberQ[b1, a3] || MemberQ[b3, a1];
  interact[{2, a2_, x2_, b2_, r_}, _,
          {4, a4_, x4_, b4_, _}, _, _] :=
              {2, a2, x2, b2, r + W} /;
                  MemberQ[b2, a4] || MemberQ[b4, a2];
```

```
interact[{3, a3_, x3_, b3_, r_}, _, _,
         {1, a1_, x1_, b1_, _}, _] :=
              {3, a3, x3, b3, r + W} /;
                    MemberQ[b3, a1] || MemberQ[b1, a3];
interact[{4, a4_, x4_, b4_, r_}, _, _, _,
         {2, a2_, x2_, b2_, _}] :=
              {4, a4, x4, b4, r + W} /;
                    MemberQ[b4, a2] || MemberQ[b2, a4];
interact[{1, a1_, x1_, b1_, r_},
         {3, a3_, x3_, b3_, _}, _, _, _] :=
              {1, a1, x1, Union[b1, b3, Drop[{a3}, x3]],
                              r + payoff[x1, x3]};
interact[{2, a2_, x2_ ,b2_, r_}, _,
         {4, a4_, x4_, b4_, _}, _, _] :=
              {2, a2, x2, Union[b2, b4, Drop[{a4}, x4]],
                              r + payoff[x2, x4]};
interact[{3, a3_, x3_, b3_, r_}, _, _,
         {1, a1_, x1_, b1_, _}, _] :=
              {3, a3, x3, Union[b3, b1, Drop[{a1}, x1]],
                              r + payoff[x3, x1]};
interact[{4, a4_, x4_, b4_, r_}, _, _, _,
         {2, a2_, x2_, b2_, _}] :=
              {4, a4, x4, Union[b4, b2, Drop[{b2}, x2]],
                              r + payoff[x4, x2]};
interact[z_, _, _, _, _] := z;
walk[{1,a___},0,_,_,{4,___},_,_,_,_,_,_] := {RND,a};
walk[{1,a___},0,_,_,_,_,_,{2,___},_,_,_] := {RND,a};
walk[{1,a___},0,_,_,_,_,_,_,{3,___},_,_] := {RND,a};
walk[{1,a___},0,_,_,_,_,_,_,_,_,_] := 0;
walk[{2,a___},_,0,_,_,{3,___},_,_,_,_,_] := {RND,a};
walk[{2,a___},_,0,_,_,_,{1,___},_,_,_,_] := {RND,a};
walk[{2,a___},_,0,_,_,_,_,_,_,{4,___},_] := {RND,a};
walk[{2,a___},_,0,_,_,_,_,_,_,_,_] := 0;
walk[{3,a___},_,_,0,_,_,{4,___},_,_,_,_] := {RND,a};
walk[{3,a___},_,_,0,_,_,_,{2,___},_,_,_] := {RND,a};
walk[{3,a___},_,_,0,_,_,_,_,_,{1,___},_] := {RND,a};
walk[{3,a___},_,_,0,_,_,_,_,_,_,_] := 0;
walk[{4,a___},_,_,_,0,_,_,{1,___},_,_,_] := {RND,a};
walk[{4,a___},_,_,_,0,_,_,_,{3,___},_,_] := {RND,a};
walk[{4,a___},_,_,_,0,_,_,_,_,_,{2,___}] := {RND,a};
walk[{4,a___},_,_,_,0,_,_,_,_,_,_] := 0;
walk[{_,a___},_,_,_,_,_,_,_,_,_,_] := {RND,a};
walk[0,{3,___},{4,___},_,_,_,_,_,_,_,_] := 0;
walk[0,{3,___},_,{1,___},_,_,_,_,_,_,_] := 0;
walk[0,{3,___},_,_,{2,___},_,_,_,_,_,_] := 0;
walk[0,_,{4,___},{1,___},_,_,_,_,_,_,_] := 0;
walk[0,_,{4,___},_,{2,___},_,_,_,_,_,_] := 0;
walk[0,_,_,{1,___},{2,___},_,_,_,_,_,_] := 0;
walk[0,{3,a___},_,_,_,_,_,_,_,_,_] := {RND,a};
walk[0,_,{4,a___},_,_,_,_,_,_,_,_] := {RND,a};
walk[0,_,_,{1,a___},_,_,_,_,_,_,_] := {RND,a};
walk[0,_,_,_,{2,a___},_,_,_,_,_,_] := {RND,a};
walk[0,_,_,_,_,_,_,_,_,_,_] := 0;
vonNeumann[func_, lat_] :=
  MapThread[func, Map[RotateRight[lat, #]&,
        {{0, 0}, {1, 0}, {0, -1}, {-1, 0}, {0, 1}}], 2];
```

```
GN[func_, lat_] :=
  MapThread[func, Map[RotateRight[lat, #]&,
          {{0, 0}, {1, 0}, {0, -1}, {-1, 0}, {0, 1},
           {1, -1}, {-1, -1}, {-1, 1}, {1, 1}, {2, 0},
           {0, -2}, {-2, 0}, {0, 2}}], 2];
  NestList[vonNeumann[interact, GN[walk, #]]&, society, t]]
```

3.7.3 vibes

```
vibes[n_, p_, g_, {P_, R_, S_, T_, W_}, t_] :=
Module[{RND, society, decide, interact, costbenefit,
                                walk, vonNeumann, GN},
  RND := Random[Integer, {1, 4}];
  society = Table[Floor[p + Random[]], {n}, {n}] /.
           1 :> {RND, Floor[g + Random[]], Random[], 0};
  decide[{1, a_, x_, b_}, {3, _, y_, _}, _, _, _] :=
                      {1, a, x, b, Floor[y + Random[]]};
  decide[{2, a_, x_, b_}, _, {4, _, y_, _}, _, _] :=
                      {2, a, x, b, Floor[y + Random[]]};
  decide[{3, a_, x_, b_}, _, _, {1, _, y_, _}, _] :=
                      {3, a, x, b, Floor[y + Random[]]};
  decide[{4, a_, x_, b_}, _, _, _, {2, _, y_, _}] :=
                      {4, a, x, b, Floor[y + Random[]]};
  decide[z_, _, _, _, _] := z;
  costbenefit[1, 1, 1] = R;
  costbenefit[0, 1, 1] = T;
  costbenefit[1, 0, 1] = S;
  costbenefit[0, 0, 1] = P;
  costbenefit[_, _, 0] = W;
  interact[{3, a_, x_, b_, k_}, _, _,
          {1, d_, _, _, h_}, _] :=
                {3, a, x, b + costbenefit[a, d, (k * h)]};
  interact[{2, a_, x_, b_, k_}, _,
          {4, d_, _, _, h_}, _, _] :=
                {2, a, x, b + costbenefit[a, d, (k * h)]};
  interact[{1, a_, x_, b_, k_},
          {3, d_, _, _, h_}, _, _, _] :=
                {1, a, x, b + costbenefit[a, d, (k * h)]};
  interact[{4, a_, x_, b_, k_}, _, _, _,
          {2, d_, _, _, h_}] :=
                {4, a, x, b + costbenefit[a, d, (k * h)]};
  interact[z_, _, _, _, _] := z;
  walk[{1,a___},0,_,_,_,{4,___},_,_,_,_,_,_,_] := {RND,a};
  walk[{1,a___},0,_,_,_,_,_,{2,___},_,_,_,_] := {RND,a};
  walk[{1,a___},0,_,_,_,_,_,_,{3,___},_,_,_] := {RND,a};
  walk[{1,a___},0,_,_,_,_,_,_,_,_,_,_,_] := 0;
  walk[{2,a___},_,0,_,_,{3,___},_,_,_,_,_,_] := {RND,a};
  walk[{2,a___},_,0,_,_,_,{1,___},_,_,_,_,_] := {RND,a};
  walk[{2,a___},_,0,_,_,_,_,_,{4,___},_,_] := {RND,a};
  walk[{2,a___},_,0,_,_,_,_,_,_,_,_,_] := 0;
  walk[{3,a___},_,_,0,_,_,{4,___},_,_,_,_,_] := {RND,a};
  walk[{3,a___},_,_,0,_,_,_,{2,___},_,_,_,_] := {RND,a};
  walk[{3,a___},_,_,0,_,_,_,_,_,{1,___},_] := {RND,a};
  walk[{3,a___},_,_,0,_,_,_,_,_,_,_,_] := 0;
```

```
walk[{4,a___},_,_,_,0,_,_,{1,___},_,_,_,_,_] := {RND,a};
walk[{4,a___},_,_,_,0,_,_,_,{3,___},_,_,_,_] := {RND,a};
walk[{4,a___},_,_,_,0,_,_,_,_,_,_,_,{2,___}] := {RND,a};
walk[{4,a___},_,_,_,0,_,_,_,_,_,_,_,_] := 0;
walk[{_,a___},_,_,_,_,_,_,_,_,_,_,_,_] := {RND,a};
walk[0,{3,___},{4,___},_,_,_,_,_,_,_,_,_] := 0;
walk[0,{3,___},_,{1,___},_,_,_,_,_,_,_,_] := 0;
walk[0,{3,___},_,_,{2,___},_,_,_,_,_,_,_] := 0;
walk[0,_,{4,___},{1,___},_,_,_,_,_,_,_,_] := 0;
walk[0,_,{4,___},_,{2,___},_,_,_,_,_,_,_] := 0;
walk[0,_,_,{1,___},{2,___},_,_,_,_,_,_,_] := 0;
walk[0,{3,a___},_,_,_,_,_,_,_,_,_,_,_] := {RND,a};
walk[0,_,{4,a___},_,_,_,_,_,_,_,_,_,_] := {RND,a};
walk[0,_,_,{1,a___},_,_,_,_,_,_,_,_,_] := {RND,a};
walk[0,_,_,_,{2,a___},_,_,_,_,_,_,_,_] := {RND,a};
walk[0,_,_,_,_,_,_,_,_,_,_,_,_] := 0;
vonNeumann[func_, lat_] :=
  MapThread[func, Map[RotateRight[lat, #]&,
        {{0, 0}, {1, 0}, {0, -1}, {-1, 0}, {0, 1}}], 2];
GN[func_, lat_] :=
  MapThread[func, Map[RotateRight[lat, #]&,
          {{0, 0}, {1, 0}, {0, -1}, {-1, 0}, {0, 1},
           {1, -1}, {-1, -1}, {-1, 1}, {1, 1}, {2, 0},
           {0, -2}, {-2, 0}, {0, 2}}], 2];
NestList[GN[walk, vonNeumann[interact,
                vonNeumann[decide, #]]]&, society, t]];
```

4

Choosing How to Behave

"We know how to behave! We've had lessons."

—John Lennon in *A Hard Day's Night* (1964)

4.1 Introduction

The classical prisoner's dilemma model discussed in Chapter 3 assumes that there are only two types of interaction behaviors: being good all of the time or being bad all of the time. Real people, of course, are not nearly that simple-minded in choosing their interaction behavior. A person can keep track in his mind of his own past interaction behavior and the result of that behavior, as well as the past interaction behavior of others. This enables a person to learn from this past interaction history and to adapt his behavior accordingly. Two models in Chapter 3 examined the easiest adaptation that a person can make based on the past, which is to simply refuse to interact with anyone who has treated him, or someone he knows, badly before. There are numerous other, more sophisticated adaptations that can be made based on a person's memory of past behavior. For example, a person could interact with anyone regardless of past behavior, but tailor his behavior based on that person's past history. Another response could involve introspection in which a person monitors his own interaction history and adjusts his behavior based on how well his past behavior has *paid off* for him. In this chapter, we look at models of each of these behavioral strategies and their effect on an individual's resource levels over time.

4.2 Choosing One's Interaction Behavior with Another Individual Based on the Behavioral History of the Other Individual

The ability of people to recognize people and remember past interaction behavior are prerequisites for engaging in *reciprocity*. We develop a model in which each person keeps a record of the past behaviors of every individual with whom he has interacted, thereby enabling him to choose how to behave with a particular individual using various criteria based on previous encounters.

In order to demonstrate how this model works, we consider a very simple system [Stebbins 1996; Rees 1997] comprised of individuals using one of the four behavior strategies:

- always behave nicely (an individual following this strategy is called a *pollyanna*);

- always behave badly (an individual following this strategy is called a *sociopath*);

- behave nicely in the first encounter with a person and then behave in the same manner as that person did in the previous encounter (an individual following this version of the *tit-for-tat* strategy is called a *nice retaliator*); or

- behave badly in the first encounter with a person and then behave in the same manner as that person did in the previous encounter (an individual following this version of the *tit-for-tat* strategy is called a *mean retaliator*).

Note: The *responding-in-kind* behavior used in the tit-for tat strategies can be succinctly expressed in the somewhat less than golden rule *do unto another as they have done unto you*.

4.2.1 The System

Our model uses a square n by n lattice with wraparound boundary conditions. There is a population density p of individuals occupying lattice sites and the remaining sites are empty. The system evolves over a given number of time steps q.

4.2.2 Populating Society

The value of an empty site is 0.

The value of a site occupied by an individual is a five-tuple $\{a, b, c, d, e\}$ where

a — an integer value between 1 and 4, indicating the direction (N, E, S, and W) faced by the individual.

b — a positive integer value representing the individual's "name" (1, 2, ..., k).

c — a list containing k elements, one for each of the individuals (including himself) in the system.

Note: The position of an element in the list indicates a person's name (e.g., the fifth element is person 5). The value of an element in the list is itself a list consisting of 0s and 1s representing a record of the past behavior of that person with the individual. A value of 0 indicates bad behavior and a value of 1 indicates good behavior (e.g., if the fifth element in the individual's list is {0, 0, 1}, it indicates that person 5 was good in his last interaction with the individual and bad on the first and second interactions with the individual).

d — an integer value indicating the individual's resource level.

e — an integer value indicating the behavioral strategy used by the individual.

The system configuration consists of a population density p of individuals. For our simple system, a fraction s are pollyannas, a fraction t are nice retaliators, a fraction u are sociopaths, and a fraction v of the individuals are mean retaliators.

Initially, individuals are randomly distributed on the lattice with zero resources and face randomly chosen directions. The elements in the initial list of the behavioral history of all of the individuals in the society are empty lists because none of the individuals has interacted yet.

This initial configuration is created as follows.

```
RND := Random[Integer, {1, 4}];
k = 0;
society = Table[Floor[p + Random[]], {n}, {n}]  /.
   1 :→ {++k, Floor[1 + v + u + Random[]]}  /.
  {{m_, 1} :→ {RND, m, Table[{}, {k}], 0,
                      Floor[1 + t/(s + t) + Random[]]},
     {m_, 2} :→ {RND, m, Table[{}, {k}], 0,
                      Floor[3 + v/(v + u) + Random[]]}};
```

4.2.3 Executing a Time Step

The partial steps, executed consecutively in each time step, are as follows.

- An individual decides how to behave in a face-to-face interaction with a person, based on what behavioral strategy he is following and the history of the behavior of the other person in previous interactions.

- The facing individuals interact, each changing his resource level and modifying his record of the other individual's interaction behavior accordingly.

- Each individual moves to the nearest neighbor site he faces, subject to the excluded volume constraint.

Deciding

In general, each individual facing another person will decide whether to be good or bad, based on what strategy he is using and none, some, or all of the history of the other person's previous behavior.

The decide rules for any strategy based on behavioral history have the general form

```
decide[{1, name1_, lis1_, res1_, strat1_},
       {3, name3_, _, _, _}, _, _, _] :=
               {1, name1, lis1, res1,
                   behave[strat1, lis1[[name3]]], strat1}
decide[{3, name3_, lis3_, res3_, strat3_}, _, _,
       {1, name1_, _, _, _}, _] :=
               {3, name3, lis3, res3,
                   behave[strat3, lis3[[name1]]], strat3}
decide[{2, name2_, lis2_, res2_, strat2_}, _,
       {4, name4_, _, _, _}, _, _] :=
               {2, name2, lis2, res2,
                   behave[strat2, lis2[[name4]]], strat2}
decide[{4, name4_, lis4_, res4_, strat4_}, _, _, _,
       {2, name2_, _, _, _}] :=
               {4, name4, lis4, res4,
                   behave[strat4, lis4[[name2]]], strat4}
```

where the behave rules depend on the specific strategies being used. For a society of people practicing simple strategies such as pollyannas (strategy 1), nice retaliators (strategy 2), sociopaths (strategy 3), and mean retaliators (strategy 4), the behave rules are:

```
behave[1, _] = 1
behave[2, {___, 0}] = 0
behave[2, _] = 1
behave[3, _] = 0
behave[4, {___, 1}] = 1
behave[4, _] = 0
```

Another decide rule leaves other individuals and empty sites unchanged.

```
decide[z_, _, _, _, _] := z
```

The decide rules are applied to the lattice sites using the anonymous function

```
vonNeumann[decide, #]&
```

where

```
vonNeumann[func_, lat_] :=
  MapThread[func, Map[RotateRight[lat, #]&,
          {{0, 0}, {1, 0}, {0, -1}, {-1, 0}, {0, 1}}], 2]
```

Interacting

In general, each individual, having decided what behavior to use in an interaction, carries out the interaction. He changes his resource level accordingly and memorizes how the other individual has behaved.

```
act[{1, name1_, lis1_, res1_, behave1_, strat1_},
    {3, name3_, _, _, behave3_, _}, _, _, _] :=
            {1, name1, ReplacePart[lis1,
              Join[lis1[[name3]], {behave3}], name3],
              res1 + payoff[behave1, behave3], strat1}

act[{3, name3_, lis3_, res3_, behave3_, strat3_}, _, _,
    {1, name1_, _, _, behave1_, _}, _] :=
            {3, name3, ReplacePart[lis3,
              Join[lis3[[name1]], {behave1}], name1],
              res3 + payoff[behave3, behave1], strat3}

act[{2, name2_, lis2_, res2_, behave2_, strat2_}, _,
    {4, name4_, _, _, behave4_, _}, _, _] :=
            {2, name2, ReplacePart[lis2,
              Join[lis2[[name4]], {behave4}], name4],
              res2 + payoff[behave2, behave4], strat2}

act[{4, name4_, lis4_, res4_, behave4_, strat4_}, _, _,
    _, {2, name2_, _, _, behave2_, _}] :=
            {4, name4, ReplacePart[lis4,
              Join[lis4[[name2]], {behave2}], name2],
              res4 + payoff[behave4, behave2], strat4}
```

where the payoff rules are given by

```
payoff[1, 1] = R
payoff[0, 1] = T
payoff[1, 0] = S
payoff[0, 0] = P
```

Another act rule leaves other individuals and empty sites unchanged.

```
act[z_, _, _, _, _] := z
```

The act rules are applied to the lattice sites using the anonymous function

```
vonNeumann[act, #]&
```

Moving

The 28 rules for the movement of the individuals are given by the set of walk rules in Chapter 1.

The walk rules are applied to the lattice sites using the anonymous function

```
GN[walk, vonNeumann[act, vonNeumann[decide, #]]]&
```

where

```
GN[func_, lat_] :=
  MapThread[func, Map[RotateRight[lat, #]&,
            {{0, 0}, {1, 0}, {0, -1}, {-1, 0}, {0, 1},
             {1, -1}, {-1, -1}, {-1, 1}, {1, 1}, {2, 0},
             {0, -2}, {-2, 0}, {0, 2}}], 2]
```

4.2.4 Evolving the System

The system evolves over q time steps, starting with the initial lattice configuration, society, using the following nesting operation.

```
NestList[GN[walk,
    vonNeumann[act, vonNeumann[decide, #]]]&, society, q]
```

4.2.5 The Program

n = lattice size
p = population density
$\{s, t, u, v\}$ = probability of being pollyanna, nice retaliator, sociopath, or mean retaliator
$\{P, R, S, T\}$ = payoff matrix entries
q = time steps

```
pollysociotit4tat[n_, p_, {s_, t_, u_, v_},
                          {P_, R_, S_, T_}, q_] :=
Module[{RND, k = 0, society, behave, decide,
                payoff, act, walk, vonNeumann, GN},

  RND := Random[Integer, {1, 4}];
  society = Table[Floor[p + Random[]], {n}, {n}]  /.
    1 :> {++k, Floor[1 + v + u + Random[]]} /.
    {{m_, 1} :> {RND, m, Table[{}, {k}], 0,
                    Floor[1 + t/(s + t) + Random[]]},
     {m_, 2} :> {RND, m, Table[{}, {k}], 0,
                    Floor[3 + v/(v + u) + Random[]]}};
  behave[1, _] = 1;
  behave[2, {___, 0}] = 0;
  behave[2, _] = 1;
  behave[3, _] = 0;
  behave[4, {___, 1}] = 1;
  behave[4, _] = 0;

  decide[{1, name1_, lis1_, res1_, strat1_},
         {3, name3_, _, _, _}, _, _, _] :=
           {1, name1, lis1, res1,
              behave[strat1, lis1[[name3]]], strat1};
```

```
decide[{3, name3_, lis3_, res3_, strat3_}, _, _,
       {1, name1_, _, _, _}, _] :=
              {3, name3, lis3, res3,
                 behave[strat3, lis3[[name1]]], strat3};
decide[{2, name2_, lis2_, res2_, strat2_}, _,
       {4, name4_, _, _, _}, _, _] :=
              {2, name2, lis2, res2,
                 behave[strat2, lis2[[name4]]], strat2};
decide[{4, name4_, lis4_, res4_, strat4_}, _, _, _,
       {2, name2_, _, _, _}] :=
              {4, name4, lis4, res4,
                 behave[strat4, lis4[[name2]]], strat4};
decide[z_, _, _, _, _] := z;

payoff[1, 1] = R;
payoff[0, 1] = T;
payoff[1, 0] = S;
payoff[0, 0] = P;

act[{1, name1_, lis1_, res1_, behave1_, strat1_},
    {3, name3_, _, _, behave3_, _}, _, _, _] :=
           {1, name1, ReplacePart[lis1,
              Join[lis1[[name3]], {behave3}], name3],
              res1 + payoff[behave1, behave3], strat1};

act[{3, name3_, lis3_, res3_, behave3_, strat3_}, _, _,
    {1, name1_, _, _, behave1_, _}, _] :=
           {3, name3, ReplacePart[lis3,
              Join[lis3[[name1]], {behave1}], name1],
              res3 + payoff[behave3, behave1], strat3};

act[{2, name2_, lis2_, res2_, behave2_, strat2_}, _,
    {4, name4_, _, _, behave4_, _}, _, _] :=
           {2, name2, ReplacePart[lis2,
              Join[lis2[[name4]], {behave4}], name4],
              res2 + payoff[behave2, behave4], strat2};

act[{4, name4_, lis4_, res4_, behave4_, strat4_}, _, _,
    _, {2, name2_, _, _, behave2_, _}] :=
           {4, name4, ReplacePart[lis4,
              Join[lis4[[name2]], {behave2}], name2],
              res4 + payoff[behave4, behave2], strat4};
act[z_, _, _, _, _] := z;

(* walk rules go here *)

vonNeumann[func_, lat_] :=
  MapThread[func, Map[RotateRight[lat, #]&,
        {{0, 0}, {1, 0}, {0, -1}, {-1, 0}, {0, 1}}], 2];

GN[func_, lat_] :=
  MapThread[func, Map[RotateRight[lat, #]&,
          {{0, 0}, {1, 0}, {0, -1}, {-1, 0}, {0, 1},
           {1, -1}, {-1, -1}, {-1, 1}, {1, 1}, {2, 0},
           {0, -2}, {-2, 0}, {0, 2}}], 2];
```

```
NestList[GN[walk,
    vonNeumann[act, vonNeumann[decide, #]]]&, society, q]]
```

4.2.6 Running the Simulation

We run the pollysociotit4tat program on a 50 by 50 lattice having a 70% population density equally divided between the four types of individuals (pollyannas, sociopaths, mean retaliators, and nice retaliators), and payoff values of {-1, 1, -2, 2} over 500 time steps

```
SeedRandom[19]
results = pollysociotit4tat[50, 0.7,
            {0.25, 0.25, 0.25, 0.25,}, {-1, 1, -2, 2}, 500];
```

For the lattice configuration at a particular time step, the following avgResourceLevel function can be used to compute the average resource level of a given strategy for that time step.

```
avgResourceLevel[lat_, strat_] :=
(Apply[Plus, #]/Length[#])&[
   Cases[lat, {_, _, _, x_, strat} :> x, 2] /. {} → {0}]
```

The avgResourceLevel function can be used to compute the average resource level of a particular strategy for each time step.

```
resourceLevels =
  Table[Map[avgResourceLevel[#, i]&, results], {i, 1, 4}];
```

We can create a graphic of the resource levels of the four types of people over time, using

```
<<Graphics`;
MultipleListPlot[Sequence @@ resourceLevels,
   SymbolShape → None,
   PlotJoined → True,
   PlotLegend → {"Pollyanna", "Nice retaliator",
                 "Sociopath", "Mean retaliator"},
   AxesLabel → {"Time", "Resource Level"},
   LegendPosition → {-1, -1},
   LegendShadow → {-.05, -.05}];
```

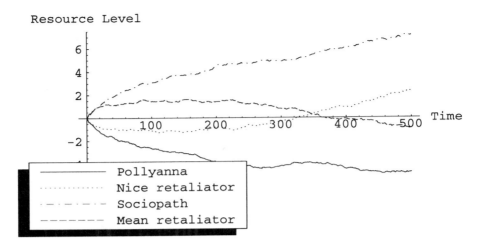

The preceding plots indicate that the sociopaths do quite well, while the pollyannas fare poorly. Stebbins [1996] has suggested that the reason for this result is that "the pollyannas [are] keeping the sociopaths alive! When there are pollyannas in the population, the sociopaths get large scores quickly, even though they beat up on each other. This gives the retaliators no chance to invade." Stebbins suggests that removing the pollyannas from the system changes the results significantly, allowing the nice retaliators to do best. Following Stebbins' suggestion of removing all of the pollyannas from the system, we found that using the Stebbins payoff values of {-10, 5, -8, 8} does indeed allow the nice retaliators to prevail but returning to the use of the payoff values of {-1, 1, -2, 2}, which produced the preceding graphic, the sociopaths still thrive at the expense of the rest of the population.

In our view, it is less than satisfying that the individuals in this model never learn from their experiences, continuing to employ the same strategy over and over again, regardless of how well they are doing (at best, there is a primitive kind of learning for the retaliators in the model, using a rather short-term memory that extends only to the previous encounter with an individual). We therefore looked at a more *introspective* model that allows an individual to change his or her behavior over the course of time based on whether a personally chosen goal is being achieved.

4.3 Aspirations

Posch [1997] has proposed a model that differs fundamentally from the previous model in that it allows a person to learn and adapt his behavior based on both his past experience and personal goals. This is done by having each individual periodically evaluate his performance by looking at his recent payoffs to determine if they meet his payoff goal and then keeping or changing his behavior based on whether he is achieving his aspiration level (this is known as *satiation*).

4.3.1 The System

Our model uses a square *n* by *n* lattice with wraparound boundary conditions. There is a population density *p* of individuals occupying lattice sites and the remaining sites stay empty. The system evolves over a given number of time steps *t*.

4.3.2 Populating Society

The value of an empty site is 0.

The value of a site occupied by an individual is a six-tuple {*a*, *b*, *c*, *d*, *e*, *f*} where the list elements are:

a — an integer value between 1 and 4, indicating the direction (N, E, S, and W) faced by the individual;

b — 0 or 1, indicating that the individual's behavior is bad (value 0) or good (value 1);

c — a real number representing the individual's aspiration level (the minimum payoff the individual must receive in order to not change behavior);

d — a real number between 0 and 1 representing the probability that the individual will evaluate whether to change behavior;

e — a list of the payoffs the individual has received since the list was last checked; and

f — an integer value indicating the individual's resource level (the sum of the payoffs from previous time steps).

The initial system population contains a fraction *g* of good guys and a fraction (1 - *g*) of bad guys. They are randomly distributed on the lattice and face randomly chosen directions. Individuals have randomly chosen aspiration levels, randomly chosen behavior evaluation probabilities, zero resources, and empty payoff lists. This configuration is created as follows.

```
RND := Random[Integer, {1, 4}]
society = Table[Floor[p + Random[]], {n}, {n}] /.
    1 :> {RND, Floor[g + Random[]],
        Random[Real, {Min[#], Max[#]}&[{R, S, T, P}]],
        Random[], {}, 0}
```

4.3.3 Executing a Time Step

The following processes occur consecutively during each time step.

- Each individual who is face-to-face with another person interacts with that person. The payoff from the interaction is added to the individual's resource level and is also placed in the individual's payoff list. All other individuals remain unchanged.

- Each individual evaluates, with a probability given by his evaluation probability value, the average payoff value in his payoff list. If this value is at least as great as the individual's aspiration level value, he empties his payoff list. If this value is less than the individual's aspiration level value, he empties his payoff list and changes his behavior.

- Each individual moves to the nearest neighbor site he is facing, subject to the excluded volume constraint.

Interacting

The rules for interaction are as follows.

```
interact[{1, b1_, a1_, e1_, p1_, r1_},
         {3, b3_, _, _, _, _}, _, _, _] :=
             {1, b1, a1, e1, Join[p1, {payoff[b1, b3]}],
                                         r1 + payoff[b1, b3]}
interact[{2, b2_, a2_ ,e2_, p2_, r2_}, _,
         {4, b4_, _, _, _, _}, _, _] :=
             {2, b2, a2, e2, Join[p2, {payoff[b2, b4]}],
                                         r2 + payoff[b2, b4]}
interact[{3, b3_, a3_, e3_, p3_, r3_}, _, _,
         {1, b1_, _, _, _, _}, _] :=
             {3, b3, a3, e3, Join[p3, {payoff[b3, b1]}],
                                         r3 + payoff[b3, b1]}
interact[{4, b4_, a4_, e4_, p4_, r4_}, _, _, _,
         {2, b2_, _, _, _, _}] :=
             {4, b4, a4, e4, Join[p4, {payoff[b4, b2]}],
                                         r4 + payoff[b4, b2]}
```

where the payoff rules are given by

```
payoff[1, 1] = R
payoff[0, 1] = T
payoff[1, 0] = S
payoff[0, 0] = P
```

The last interact rule leaves the individuals unchanged.

```
interact[z_, _, _, _, _] := z
```

The interact rules are applied to the lattice using the anonymous function

```
vonNeumann[interact, #]&
```

where

```
vonNeumann[func_, lat_] :=
  MapThread[func, Map[RotateRight[lat, #]&,
          {{0, 0}, {1, 0}, {0, -1}, {-1, 0}, {0, 1}}], 2]
```

Evaluating Aspirations

The rules for the evaluation of an individual's behavior are:

```
eval[{x_, b_, 0, e_, pp_, r_}]  :=  {x, b, 0, e, {}, r}
eval[{x_, b_, a_, e_, {}, r_}]  :=  {x, b, a, e, {}, r}
eval[{x_, 1, a_, e_, pp_, r_}]  :=  {x,
    Sign[Floor[(Apply[Plus, pp]/Length[pp])/a]],
                    a, e, {}, r} /; (Random[] + e) > 1
eval[{x_, 0, a_, e_, pp_, r_}]  :=  {x,
    1 - Sign[Floor[(Apply[Plus, pp]/Length[pp])/a]],
                    a, e, {}, r} /; (Random[] + e) > 1
eval[z_, _, _, _, _]  :=  z
```

The first eval rule says that if an individual has a zero aspiration level, then he never changes his behavior.

The second eval rule says that if an individual has not interacted with anyone since his last evaluation, he doesn't change his behavior.

The third and fourth eval rules say that if an individual has a nonzero aspiration value and has interacted at least once since his last evaluation, then with probability e the individual compares his average payoff since his previous evaluation with his aspiration level. The average payoff is calculated using Apply[Plus, p]/Length[p]. If the average payoff meets or exceeds the individual's aspiration level a, he continues to follow the behavior he has been using. If the average payoff is less than the aspiration level, the individual changes his behavior. In either case, the individual empties his payoff list.

The fifth eval rule leaves all other individuals and empty sites unchanged.

The eval rules are applied to the lattice using the anonymous function

```
Map[eval, vonNeumann[interact, #], {2}]&
```

Moving

The 28 rules for the movement of the individual are given by the set of walk rules in Chapter 1.

The walk rules are applied to the lattice sites using the anonymous function

```
GN[walk, Map[eval, vonNeumann[interact, #], {2}]]&
```

where

```
GN[func_, lat_] :=
  MapThread[func, Map[RotateRight[lat, #]&,
            {{0, 0}, {1, 0}, {0, -1}, {-1, 0}, {0, 1},
             {1, -1}, {-1, -1}, {-1, 1}, {1, 1}, {2, 0},
             {0, -2}, {-2, 0}, {0, 2}}], 2]
```

4.3.4 Evolving the System

The system evolves over t time steps, starting with the initial lattice configuration, society, using the following nesting operation.

```
NestList[GN[walk,
    Map[eval, vonNeumann[interact, #], {2}]]&, society, t]
```

4.3.5 The Program

```
aspirations[n_, p_, g_, {P_, R_, S_, T_}, t_] :=
Module[{society, interact, payoff,  eval, walk, vonNeumann,
GN, RND},

  RND := Random[Integer, {1, 4}];
  society = Table[Floor[p + Random[]], {n}, {n}] /.
    1 :> {RND, Floor[g + Random[]],
          Random[Real, {Min[#], Max[#]}&[{R, S, T, P}]],
          Random[], {}, 0};

  payoff[1, 1] = R;
  payoff[0, 1] = T;
  payoff[1, 0] = S;
  payoff[0, 0] = P;

  interact[{1, b1_, a1_, e1_, p1_, r1_},
          {3, b3_, _, _, _, _}, _, _, _] :=
            {1, b1, a1, e1, Join[p1, {payoff[b1, b3]}],
                            r1 + payoff[b1, b3]};
  interact[{2, b2_, a2_ ,e2_, p2_, r2_}, _,
          {4, b4_, _, _, _, _}, _, _] :=
            {2, b2, a2, e2, Join[p2, {payoff[b2, b4]}],
                            r2 + payoff[b2, b4]};
  interact[{3, b3_, a3_, e3_, p3_, r3_}, _, _,
          {1, b1_, _, _, _, _}, _] :=
            {3, b3, a3, e3, Join[p3, {payoff[b3, b1]}],
                            r3 + payoff[b3, b1]};
  interact[{4, b4_, a4_, e4_, p4_, r4_}, _, _, _,
          {2, b2_, _, _, _, _}] :=
            {4, b4, a4, e4, Join[p4, {payoff[b4, b2]}],
                            r4 + payoff[b4, b2]};
  interact[z_, _, _, _, _] := z;
```

```
eval[{x_, b_, 0, e_, pp_, r_}] := {x, b, 0, e, {}, r};
eval[{x_, b_, a_, e_, {}, r_}] := {x, b, a, e, {}, r};
eval[{x_, 1, a_, e_, pp_, r_}] :=  {x,
  Sign[Floor[(Apply[Plus, pp]/Length[pp])/a]],
                      a, e, {}, r} /; (Random[] + e) > 1;
eval[{x_, 0, a_, e_, pp_, r_}] :=  {x,
  1 - Sign[Floor[(Apply[Plus, pp]/Length[pp])/a]],
                      a, e, {}, r} /; (Random[] + e) > 1;
eval[z_, _, _, _, _] := z;

(* walk rules go here *)

vonNeumann[func_, lat_] :=
  MapThread[func, Map[RotateRight[lat, #]&,
        {{0, 0}, {1, 0}, {0, -1}, {-1, 0}, {0, 1}}], 2];

GN[func_, lat_] :=
  MapThread[func, Map[RotateRight[lat, #]&,
        {{0, 0}, {1, 0}, {0, -1}, {-1, 0}, {0, 1},
         {1, -1}, {-1, -1}, {-1, 1}, {1, 1}, {2, 0},
         {0, -2}, {-2, 0}, {0, 2}}], 2];

NestList[GN[walk,
  Map[eval, vonNeumann[interact, #], {2}]]&, society, t]]
```

4.3.6 Running the Simulation

We run the aspirations program on a 25 by 25 lattice having a 60% population density equally divided between good guys and bad guys and payoff values of {1, 3, 0, 5} over 200 time steps.

```
SeedRandom[23];
results = aspirations[25, 0.6, 0.5, {1, 3, 0, 5}, 200];
```

The following two graphics plot resource level with aspiration level and with the probability of re-evaluation.

```
ListPlot[
  Cases[Last[results], {_, _, a_, _, _, r_} :> {a, r}, 2],
  AxesLabel → {"Aspiration", "Resources"}];

ListPlot[
  Cases[Last[results], {_, _, _, e_, _, r_} :> {e, r}, 2],
  AxesLabel → {"Evaluation %","Resources"}];
```

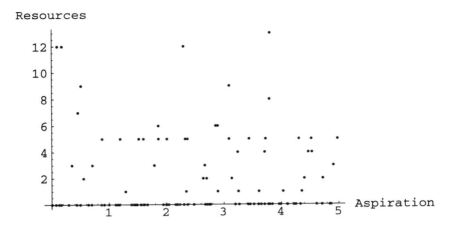

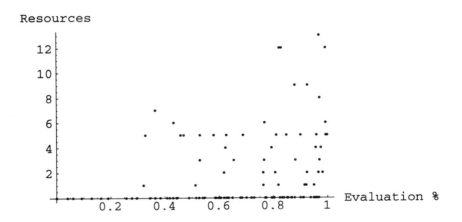

The plots indicate that aspiration level is not connected to the ability to accumulate resources but that the likelihood of strategy re-evaluation is correlated to resource accumulation. In particular, those people who are more likely to re-evaluate whether their strategy is performing well appear more likely to receive a higher payoff over time (so that introspection pays off).

4.4 Computer Simulation Projects

1. A variant of the tit-for-tat strategy is the tit-for-two-tats strategy in which an individual doesn't retaliate by behaving badly against another person until that person has behaved badly in two successive interactions. Incorporate this strategy into the behavioral history-based strategy program by modifying the behave rules.

2. In one variation of the aspiration model, each individual's probability of re-evaluating his behavior is not fixed but, instead, increases or decreases with the difference between his last payoff and his aspiration level. This can be implemented by changing the interact rules to

```
interact[{1, b1_, a1_, e1_, p1_, r1_},
         {3, b3_, _, _, _, _}, _, _, _] :=
            {1, b1, a1, e1 - payoff[b1, b3], Join[
             p1, {payoff[b1, b3]}], r1 + payoff[b1, b3]}
interact[{2, b2_, a2_ ,e2_, p2_, r2_}, _,
         {4, b4_, _, _, _, _}, _, _] :=
            {2, b2, a2, e2 - payoff[b2, b4], Join[
             p2, {payoff[b2, b4]}], r2 + payoff[b2, b4]}
interact[{3, b3_, a3_, e3_, p3_, r3_}, _, _,
         {1, b1_, _, _, _, _}, _] :=
            {3, b3, a3, e3 - payoff[b3, b1], Join[
             p3, {payoff[b3, b1]}], r3 + payoff[b3, b1]}
interact[{4, b4_, a4_, e4_, p4_, r4_}, _, _, _,
         {2, b2_, _, _, _, _}] :=
            {4, b4, a4, e4 - payoff[b4, b2], Join[
             p4, {payoff[b4, b2]}], r4 + payoff[b4, b2]}
```

Implement this code in a program and run it to see what happens.

3. An individual could keep a history of his own behavior and payoffs and bias his choice of behavior towards the behavior producing the higher payoff. For example, a person would be twice as likely to choose one behavior over another if the average payoff from that behavior had been twice as much as the average payoff from the other behavior. Implement this strategy and run the simulation to see how the resource levels are affected.

4. In yet another variation of the aspiration model, the payoffs for an interaction between any two facing individuals varies. For each interaction that takes place, the values of R, S, T, and P are chosen randomly using Random[Real, $\{\alpha, \beta\}$] where α and β are integers (e.g., $\alpha = -3$ and $\beta = 5$). Implement this code in a program and run it to see what happens.

5. Nowak and co-workers [Nowak et al. 1994; 1995] have developed a *keeping up with the Joneses* version of the win stay-lose shift strategy. All of the lattice sites are occupied by individuals (who are therefore immobile) who interact with each of their nearest neighbors over a number of time steps. The interaction behavior (cooperate or defect) used by a person in each time step is determined as follows: In a given time step, a person interacts with himself and with his eight nearest neighbors (ie., the nine sites in the Moore neighborhood of the person) in a pairwise manner with either good or bad behavior. The total payoff to the person resulting from all nine pairwise interactions is determined. The person then compares his total payoff to the payoffs received by all of the people in his neighborhood (who similarly have interacted with their neighbors), and adapts for the next time step the strategy of the neighbor (which included himself) who received the biggest payoff.

This *static* spatial prisoner's dilemma can be modified to incorporate the mobility of the agents, thereby enabling interaction partners to change over time. We present the development of this model here in detail because it contains some interesting *Mathematica* programming.

The model employs an n by n square lattice, with periodic boundary conditions. The lattice sites are either empty or occupied. The site values are integer values.

The value of this single element is the integer 0, 1, or -1 where the values have the meanings:

0 — a site occupied by a bad guy (defector or noncooperator);

1 — a site occupied by a good guy (cooperator); and

-1 — an empty site.

Note: Until now, we have used 0 to denote an empty site. The usefulness of letting -1 represent empty sites in this model becomes apparent in the following description of the payoff rules.

For a density r of people, of whom a fraction k are bad guys and the rest are good guys, the initial configuration is given by

```
society = Table[Floor[-1 + r + Random[]], {n}, {n}] /.
            0 :> Floor[1 - k + Random[]]
```

The prisoner's dilemma "game" is played over a number of time steps (i.e., rounds). In each time step, the following occur consecutively.

- Each site on the lattice interacts in a pairwise manner with itself and with the eight nearest neighbor sites in its Moore neighborhood and the total payoff to each site from these interactions is calculated.

- Each site examines the total payoff received by each site in its neighborhood. If the site is empty it remains empty. If the site is occupied by a good guy or a bad guy, he adopts the strategy (remains good or bad or *switches sides* from good to bad or vice versa) based on the strategy used by the person in his neighborhood that received the largest payoff, and randomly chooses a direction (north, east, south, or west) to face.

- Each person on an occupied site that faces an empty site not faced by any other person moves to that site. All other individuals remain in place. An empty site faced by exactly one person becomes occupied by that person. All other empty sites remain unoccupied.

We implement these stages of each time step as follows.

Calculating the {payoff to site, strategy of site} matrix

Each site interacts with the sites in its Moore neighborhood in a pairwise manner. A payoff value is computed for each interaction as follows.

- An empty site interacting with a good guy or a bad guy or an empty site has a payoff of 0.

- A good guy interacting with a good guy has a payoff of 1.

- A good guy interacting with a bad guy has a payoff of 0.

- A good guy interacting with an empty site has a payoff of 0.

- A bad guy interacting with a defector has a payoff of 0.

- A bad guy interacting with a good guy has a payoff of p (where $p > 1.0$).

- A bad guy interacting with an empty site has a payoff of 0.

Note: In the absence of empty sites ($r = 0$), these payoff values correspond to using $S = P = 0$, $R = 1$, $T = p$ ($p > 1$) in the static prisoner's dilemma game. These values are used to simplify the program by using a single payoff parameter value but other payoff values should produce the same results.

The total payoff for a site is the sum of the nine payoffs resulting from the interactions of the site with each site in its neighborhood (including itself). The total payoff to an empty site is zero.

Note: In the payoff calculation that follows, an empty site (i.e., one having a value of -1) is treated as a bad guy (with a value of 0) and so the total payoff to an empty site may be incorrectly calculated as nonzero, but this is irrelevant since empty sites are dealt with differently than occupied sites in the later steps of the calculation.

We take the matrix whose elements are integer-number values representing the strategies of the sites in the lattice (0 for a bad guy, 1 for a good guy, and -1 for an empty site) and use it to create a matrix whose elements are ordered pairs whose first component is the total payoff to the corresponding site in the lattice and whose second component is the strategy of the corresponding site in the lattice.

The {payoff, strategy} matrix is calculated by applying the following anonymous function to the lattice, represented here by mat.

```
MapThread[List, {Moore[payoff, (# /. -1 → 0)], #}, 2]&[mat]
```

where

```
Moore[func_, lat_] :=
  MapThread[func, Map[RotateRight[lat, #]&,
          {{0, 0}, {1, 0}, {0, -1}, {-1, 0}, {0, 1},
           {1, -1}, {-1, -1}, {-1, 1}, {1, 1}}], 2]
```

and the payoff rules are given by

```
payoff[1, a_, b_, c_, d_, e_, f_, g_, h_] :=
                    1 + a + b + c + d + e + f + g + h
payoff[0, a_, b_, c_, d_, e_, f_, g_, h_] :=
                    p (a + b + c + d + e + f + g + h)
```

where the arguments to the payoff rules are the values of the nine sites in the Moore neighborhood of a site.

Calculating the {direction faced by site, strategy of site} matrix

We take the matrix whose elements are ordered pairs whose components are the total payoff and the strategy of the corresponding site in the lattice and use it to create a matrix of elements that are ordered pairs whose components are the direction faced and the strategy adopted by the corresponding site in the lattice (i.e., the strategy used by the site in the neighborhood that received the highest payoff).

The direction faced by a site is given by an integer ranging from 0 to 4 where the values have the meanings:

0 — an empty site;

1 — a site occupied by a north facing individual;

2 — a site occupied by a east facing individual;

3 — a site occupied by a south facing individual; and

4 — a site occupied by a west facing individual.

The {direction faced, strategy used} matrix is determined using

```
Moore[strategy, MapThread[List,
          {Moore[payoff, (# /. -1 → 0)], #}, 2]]&[mat]
```

where the strategy rules are given by:

```
strategy[{_, -1}, ___] := {0, -1}
strategy[x: {_, _}..] := {Random[Integer, {1, 4}],
    Cases[#, {Max[Transpose[#][[1]]], _}][[1, 2]]&[
                        DeleteCases[{x},{_, -1}]]}
```

We take the matrix with elements of ordered pair values whose components are the direction faced and the strategy adopted by the corresponding site in the lattice and use the following rules to move the good guys and bad guys.

```
move[{1,x_},{y_?Positive,_},_,_,_,_,_,_,_,_,_,_] := x
move[{2,x_},_,{y_?Positive,_},_,_,_,_,_,_,_,_,_] := x
move[{3,x_},_,_,{y_?Positive,_},_,_,_,_,_,_,_,_] := x
move[{4,x_},_,_,_,{y_?Positive,_},_,_,_,_,_,_,_] := x
move[{1,x_},{0,-1},_,_,_,{4,_},_,_,_,_,_,_] := x
move[{1,x_},{0,-1},_,_,_,_,_,{2,_},_,_,_,_] := x
```

```
move[{1,x_},{0,-1},_,_,_,_,_,_,_,{3,_},_,_,_] := x
move[{2,x_},_,{0,-1},_,_,{3,_},_,_,_,_,_,_] := x
move[{2,x_},_,{0,-1},_,_,_,{1,_},_,_,_,_,_] := x
move[{2,x_},_,{0,-1},_,_,_,_,_,_,{4,_},_,_] := x
move[{3,x_},_,_,{0,-1},_,_,{4,_},_,_,_,_,_] := x
move[{3,x_},_,_,{0,-1},_,_,_,{2,_},_,_,_,_] := x
move[{3,x_},_,_,{0,-1},_,_,_,_,_,_,{1,_},_] := x
move[{4,x_},_,_,_,{0,-1},_,_,{1,_},_,_,_,_] := x
move[{4,x_},_,_,_,{0,-1},_,_,_,{3,_},_,_,_] := x
move[{4,x_},_,_,_,{0,-1},_,_,_,_,_,_,{2,_}] := x
move[{1,_},{0,-1},_,_,_,_,_,_,_,_,_,_,_] := -1
move[{2,_},_,{0,-1},_,_,_,_,_,_,_,_,_,_] := -1
move[{3,_},_,_,{0,-1},_,_,_,_,_,_,_,_,_] := -1
move[{4,_},_,_,_,{0,-1},_,_,_,_,_,_,_,_] := -1
move[{0,-1},{3,_},{4,_},_,_,_,_,_,_,_,_,_,_] := -1
move[{0,-1},{3,_},_,{1,_},_,_,_,_,_,_,_,_,_] := -1
move[{0,-1},{3,_},_,_,{2,_},_,_,_,_,_,_,_,_] := -1
move[{0,-1},_,{4,_},{1,_},_,_,_,_,_,_,_,_,_] := -1
move[{0,-1},_,{4,_},_,{2,_},_,_,_,_,_,_,_,_] := -1
move[{0,-1},_,_,{1,_},{2,_},_,_,_,_,_,_,_,_] := -1
move[{0,-1},{3,x_},_,_,_,_,_,_,_,_,_,_,_] := x
move[{0,-1},_,{4,x_},_,_,_,_,_,_,_,_,_,_] := x
move[{0,-1},_,_,{1,x_},_,_,_,_,_,_,_,_,_] := x
move[{0,-1},_,_,_,{2,x_},_,_,_,_,_,_,_,_] := x
move[{0,-1},_,_,_,_,_,_,_,_,_,_,_,_] := -1
```

The lattice whose site values consist of the new strategy matrix is calculated by applying the move rules to the {direction faced, strategy used} matrix using

```
GN[move, Moore[strategy, MapThread[List,
        {Moore[payoff, (# /. -1 → 0)], #}, 2]]]&[mat]
```

where

```
GN[func_, lat_] :=
  MapThread[func, Map[RotateRight[lat, #]&,
        {{0, 0}, {1, 0}, {0, -1}, {-1, 0}, {0, 1},
         {1, -1}, {-1, -1}, {-1, 1}, {1, 1}, {2, 0},
         {0, -2}, {-2, 0}, {0, 2}}], 2]
```

The mobile prisoner's dilemma program evolves over t time steps, starting with the initial lattice configuration, society, using the following nesting operation.

```
NestList[GN[move, Moore[strategy, MapThread[List,
      {Moore[payoff, (# /. -1 → 0)], #}, 2]]]&, society, t]
```

The mobile prisoner's dilemma program is assembled from the previously given code fragments.

```
mobilePrisonerDilemma[n_, r_, k_, p_, t_] :=
Module[{RND, society, payoff, strategy, move, Moore, GN},

  RND := Random[Integer, {1, 4}];
  society = Table[Floor[-1 + r + Random[]], {n}, {n}] /.
        0 :> Floor[1 - k + Random[]];
```

```
payoff[1, a_, b_, c_, d_, e_, f_, g_, h_] :=
                  1 + a + b + c + d + e + f + g + h;
payoff[0, a_, b_, c_, d_, e_, f_, g_, h_] :=
                  p (a + b + c + d + e + f + g + h);

strategy[{_, -1}, ___] := {0, -1}
strategy[x:{_, _}..] := {RND,
   Cases[#, {Max[Transpose[#][[1]]], _}][[1, 2]]&[
                           DeleteCases[{x},{_, -1}]]};

move[{1,x_},{y_?Positive,_},_,_,_,_,_,_,_,_,_,_] := x;
move[{2,x_},_,{y_?Positive,_},_,_,_,_,_,_,_,_,_] := x;
move[{3,x_},_,_,{y_?Positive,_},_,_,_,_,_,_,_,_] := x;
move[{4,x_},_,_,_,{y_?Positive,_},_,_,_,_,_,_,_] := x;
move[{1,x_},{0,-1},_,_,_,{4,_},_,_,_,_,_,_] := x;
move[{1,x_},{0,-1},_,_,_,_,_,_,{2,_},_,_,_] := x;
move[{1,x_},{0,-1},_,_,_,_,_,_,_,{3,_},_,_] := x;
move[{2,x_},_,{0,-1},_,_,_,{3,_},_,_,_,_,_] := x;
move[{2,x_},_,{0,-1},_,_,_,_,_,{1,_},_,_,_] := x;
move[{2,x_},_,{0,-1},_,_,_,_,_,_,_,{4,_},_] := x;
move[{3,x_},_,_,{0,-1},_,_,{4,_},_,_,_,_,_] := x;
move[{3,x_},_,_,{0,-1},_,_,_,{2,_},_,_,_,_] := x;
move[{3,x_},_,_,{0,-1},_,_,_,_,_,_,{1,_},_] := x;
move[{4,x_},_,_,_,{0,-1},_,_,{1,_},_,_,_,_] := x;
move[{4,x_},_,_,_,{0,-1},_,_,_,{3,_},_,_,_] := x;
move[{4,x_},_,_,_,{0,-1},_,_,_,_,_,_,{2,_}] := x;
move[{1,_},{0,-1},_,_,_,_,_,_,_,_,_,_] := -1;
move[{2,_},_,{0,-1},_,_,_,_,_,_,_,_,_] := -1;
move[{3,_},_,_,{0,-1},_,_,_,_,_,_,_,_] := -1;
move[{4,_},_,_,_,{0,-1},_,_,_,_,_,_,_] := -1;
move[{0,-1},{3,_},{4,_},_,_,_,_,_,_,_,_,_] := -1;
move[{0,-1},{3,_},_,{1,_},_,_,_,_,_,_,_,_] := -1;
move[{0,-1},{3,_},_,_,{2,_},_,_,_,_,_,_,_] := -1;
move[{0,-1},_,{4,_},{1,_},_,_,_,_,_,_,_,_] := -1;
move[{0,-1},_,{4,_},_,{2,_},_,_,_,_,_,_,_] := -1;
move[{0,-1},_,_,{1,_},{2,_},_,_,_,_,_,_,_] := -1;
move[{0,-1},{3,x_},_,_,_,_,_,_,_,_,_,_] := x;
move[{0,-1},_,{4,x_},_,_,_,_,_,_,_,_,_] := x;
move[{0,-1},_,_,{1,x_},_,_,_,_,_,_,_,_] := x;
move[{0,-1},_,_,_,{2,x_},_,_,_,_,_,_,_] := x;
move[{0,-1},_,_,_,_,_,_,_,_,_,_,_] := -1;

Moore[func_, lat_] :=
  MapThread[func, Map[RotateRight[lat, #]&,
        {{0, 0}, {1, 0}, {0, -1}, {-1, 0}, {0, 1},
         {1, -1}, {-1, -1}, {-1, 1}, {1, 1}}], 2];

GN[func_, lat_] :=
  MapThread[func, Map[RotateRight[lat, #]&,
        {{0, 0}, {1, 0}, {0, -1}, {-1, 0}, {0, 1},
         {1, -1}, {-1, -1}, {-1, 1}, {1, 1}, {2, 0},
         {0, -2}, {-2, 0}, {0, 2}}], 2];

NestList[GN[move, Moore[strategy, MapThread[List,
   {Moore[payoff, (# /. -1 -> 0)], #}, 2]]]&, society, t]]
```

4.5 References

Nowak, Martin A., Sebastian Bonhoeffer, and Robert M. May. 1994. "Spatial Games and the Maintenance of Cooperation." In *Proceedings of the National Academy of Sciences USA* 91; 4877–81.

Nowak, Martin A., Robert M. May, and Karl Sigmund. 1995. "The Arithmetics of Mutual Help." *Scientific American* 272 (June): 76–81.

Posch, Martin. 1997. "Win Stay-Lose Shift: An Elementary Learning Rule for Normal Form Games." 1997 Sante Fe Institute Working Papers No. 97-06-056 E.

Rees, Peter. 1997. "Simulation Sheds Light on Cooperation." *Scientific Computing World* (June): 25–29.

Stebbins, George. 1996. "The Evolution of Cooperation as Examined Using the Prisoner's Dilemma." http://www.cs.utah.edu/~stebbins/java/cs573/paper.html

4.6 Programs in the Chapter

4.6.1 pollysociotit4tat

```
pollysociotit4tat[n_, p_, {s_, t_, u_, v_},
                         {P_, R_, S_, T_}, q_] :=
Module[{RND, k = 0, society, behave, decide,
                payoff, act, walk, vonNeumann, GN},
  RND := Random[Integer, {1, 4}];
  society = Table[Floor[p + Random[]], {n}, {n}]  /.
    1 :> {++k, Floor[1 + v + u + Random[]]} /.
    {{m_, 1} :> {RND, m, Table[{}, {k}], 0,
                      Floor[1 + t/(s + t) + Random[]]},
      {m_, 2} :> {RND, m, Table[{}, {k}], 0,
                      Floor[3 + v/(v + u) + Random[]]}};
  behave[1, _] = 1;
  behave[2, {___, 0}] = 0;
  behave[2, _] = 1;
  behave[3, _] = 0;
  behave[4, {___, 1}] = 1;
  behave[4, _] = 0;
  decide[{1, name1_, lis1_, res1_, strat1_},
         {3, name3_, _, _, _}, _, _, _] :=
                {1, name1, lis1, res1,
                    behave[strat1, lis1[[name3]]], strat1};
  decide[{3, name3_, lis3_, res3_, strat3_}, _, _,
         {1, name1_, _, _, _}, _] :=
                {3, name3, lis3, res3,
                    behave[strat3, lis3[[name1]]], strat3};
  decide[{2, name2_, lis2_, res2_, strat2_}, _,
         {4, name4_, _, _, _}, _, _] :=
                {2, name2, lis2, res2,
                    behave[strat2, lis2[[name4]]], strat2};
  decide[{4, name4_, lis4_, res4_, strat4_}, _, _, _,
         {2, name2_, _, _, _}] :=
                {4, name4, lis4, res4,
                    behave[strat4, lis4[[name2]]], strat4};
  decide[z_, _, _, _, _] := z;
  payoff[1, 1] = R;
  payoff[0, 1] = T;
  payoff[1, 0] = S;
  payoff[0, 0] = P;
  act[{1, name1_, lis1_, res1_, behave1_, strat1_},
      {3, name3_, _, _, behave3_, _}, _, _, _] :=
                {1, name1, ReplacePart[lis1,
                  Join[lis1[[name3]], {behave3}], name3],
                  res1 + payoff[behave1, behave3], strat1};

  act[{3, name3_, lis3_, res3_, behave3_, strat3_}, _, _,
      {1, name1_, _, _, behave1_, _}, _] :=
                {3, name3, ReplacePart[lis3,
                  Join[lis3[[name1]], {behave1}], name1],
```

```
                            res3 + payoff[behave3, behave1], strat3};

      act[{2, name2_, lis2_, res2_, behave2_, strat2_}, _,
         {4, name4_, _, _, behave4_, _}, _, _] :=
                {2, name2, ReplacePart[lis2,
                   Join[lis2[[name4]], {behave4}], name4],
                   res2 + payoff[behave2, behave4], strat2};

      act[{4, name4_, lis4_, res4_, behave4_, strat4_}, _, _,
         _, {2, name2_, _, _, behave2_, _}] :=
                {4, name4, ReplacePart[lis4,
                   Join[lis4[[name2]], {behave2}], name2],
                   res4 + payoff[behave4, behave2], strat4};
      act[z_, _, _, _, _] := z;
      walk[{1,a___},0,_,_,_,{4,___},_,_,_,_,_,_,_] := {RND,a};
      walk[{1,a___},0,_,_,_,_,_,_,{2,___},_,_,_,_] := {RND,a};
      walk[{1,a___},0,_,_,_,_,_,_,_,_,{3,___},_,_,_] := {RND,a};
      walk[{1,a___},0,_,_,_,_,_,_,_,_,_,_,_,_] := 0;
      walk[{2,a___},_,0,_,_,{3,___},_,_,_,_,_,_,_] := {RND,a};
      walk[{2,a___},_,0,_,_,_,{1,___},_,_,_,_,_,_] := {RND,a};
      walk[{2,a___},_,0,_,_,_,_,_,_,{4,___},_,_,_] := {RND,a};
      walk[{2,a___},_,0,_,_,_,_,_,_,_,_,_,_] := 0;
      walk[{3,a___},_,_,0,_,_,{4,___},_,_,_,_,_,_] := {RND,a};
      walk[{3,a___},_,_,0,_,_,_,{2,___},_,_,_,_,_] := {RND,a};
      walk[{3,a___},_,_,0,_,_,_,_,_,_,{1,___},_] := {RND,a};
      walk[{3,a___},_,_,0,_,_,_,_,_,_,_,_,_] := 0;
      walk[{4,a___},_,_,_,0,_,_,{1,___},_,_,_,_,_] := {RND,a};
      walk[{4,a___},_,_,_,0,_,_,_,{3,___},_,_,_,_] := {RND,a};
      walk[{4,a___},_,_,_,0,_,_,_,_,_,_,_,{2,___}] := {RND,a};
      walk[{4,a___},_,_,_,0,_,_,_,_,_,_,_,_] := 0;
      walk[{_,a___},_,_,_,_,_,_,_,_,_,_,_,_] := {RND,a};
      walk[0,{3,___},{4,___},_,_,_,_,_,_,_,_,_,_] := 0;
      walk[0,{3,___},_,{1,___},_,_,_,_,_,_,_,_,_] := 0;
      walk[0,{3,___},_,_,{2,___},_,_,_,_,_,_,_,_] := 0;
      walk[0,_,{4,___},{1,___},_,_,_,_,_,_,_,_,_] := 0;
      walk[0,_,{4,___},_,{2,___},_,_,_,_,_,_,_,_] := 0;
      walk[0,_,_,{1,___},{2,___},_,_,_,_,_,_,_,_] := 0;
      walk[0,{3,a___},_,_,_,_,_,_,_,_,_,_,_] := {RND,a};
      walk[0,_,{4,a___},_,_,_,_,_,_,_,_,_,_] := {RND,a};
      walk[0,_,_,{1,a___},_,_,_,_,_,_,_,_,_] := {RND,a};
      walk[0,_,_,_,{2,a___},_,_,_,_,_,_,_,_] := {RND,a};
      walk[0,_,_,_,_,_,_,_,_,_,_,_,_] := 0;
      vonNeumann[func_, lat_] :=
        MapThread[func, Map[RotateRight[lat, #]&,
              {{0, 0}, {1, 0}, {0, -1}, {-1, 0}, {0, 1}}], 2];
      GN[func_, lat_] :=
        MapThread[func, Map[RotateRight[lat, #]&,
              {{0, 0}, {1, 0}, {0, -1}, {-1, 0}, {0, 1},
               {1, -1}, {-1, -1}, {-1, 1}, {1, 1}, {2, 0},
               {0, -2}, {-2, 0}, {0, 2}}], 2];
      NestList[GN[walk,
        vonNeumann[act, vonNeumann[decide, #]]]&, society, q]]
```

4.6.2 aspirations

```
aspirations[n_, p_, g_, {P_, R_, S_, T_}, t_] :=
Module[{society, interact, payoff,  eval, walk, vonNeumann,
GN, RND},
   RND := Random[Integer, {1, 4}];
   society = Table[Floor[p + Random[]], {n}, {n}] /.
      1 :> {RND, Floor[g + Random[]],
            Random[Real, {Min[#], Max[#]}&[{R, S, T, P}]],
            Random[], {}, 0};
   payoff[1, 1] = R;
   payoff[0, 1] = T;
   payoff[1, 0] = S;
   payoff[0, 0] = P;
   interact[{1, b1_, a1_, e1_, p1_, r1_},
            {3, b3_, _, _, _, _}, _, _, _] :=
            {1, b1, a1, e1, Join[p1, {payoff[b1, b3]}],
                                 r1 + payoff[b1, b3]};
   interact[{2, b2_, a2_ ,e2_, p2_, r2_}, _,
            {4, b4_, _, _, _, _}, _, _] :=
            {2, b2, a2, e2, Join[p2, {payoff[b2, b4]}],
                                 r2 + payoff[b2, b4]};
   interact[{3, b3_, a3_, e3_, p3_, r3_}, _, _,
            {1, b1_, _, _, _, _}, _] :=
            {3, b3, a3, e3, Join[p3, {payoff[b3, b1]}],
                                 r3 + payoff[b3, b1]};
   interact[{4, b4_, a4_, e4_, p4_, r4_}, _, _, _,
            {2, b2_, _, _, _, _}] :=
            {4, b4, a4, e4, Join[p4, {payoff[b4, b2]}],
                                 r4 + payoff[b4, b2]};
   interact[z_, _, _, _, _] := z;
   eval[{x_, b_, 0, e_, pp_, r_}] :=  {x, b, 0, e, {}, r};
   eval[{x_, b_, a_, e_, {}, r_}] :=  {x, b, a, e, {}, r};
   eval[{x_, 1, a_, e_, pp_, r_}] :=  {x,
      Sign[Floor[(Apply[Plus, pp]/Length[pp])/a]],
                        a, e, {}, r} /; (Random[] + e) > 1;
   eval[{x_, 0, a_, e_, pp_, r_}] :=  {x,
      1 - Sign[Floor[(Apply[Plus, pp]/Length[pp])/a]],
                        a, e, {}, r} /; (Random[] + e) > 1;
   eval[z_, _, _, _, _] := z;
   walk[{1,a___},0,_,_,_,{4,___},_,_,_,_,_,_] := {RND,a};
   walk[{1,a___},0,_,_,_,_,_,_,{2,___},_,_,_,_] := {RND,a};
   walk[{1,a___},0,_,_,_,_,_,_,_,_,{3,___},_,_,_] := {RND,a};
   walk[{1,a___},0,_,_,_,_,_,_,_,_,_,_,_,_] := 0;
   walk[{2,a___},_,0,_,_,{3,___},_,_,_,_,_,_] := {RND,a};
   walk[{2,a___},_,0,_,_,_,{1,___},_,_,_,_,_,_] := {RND,a};
   walk[{2,a___},_,0,_,_,_,_,_,_,_,{4,___},_,_] := {RND,a};
   walk[{2,a___},_,0,_,_,_,_,_,_,_,_,_,_,_] := 0;
   walk[{3,a___},_,_,0,_,_,{4,___},_,_,_,_,_,_] := {RND,a};
   walk[{3,a___},_,_,0,_,_,_,{2,___},_,_,_,_,_] := {RND,a};
   walk[{3,a___},_,_,0,_,_,_,_,_,_,_,{1,___},_] := {RND,a};
   walk[{3,a___},_,_,0,_,_,_,_,_,_,_,_,_,_] := 0;
   walk[{4,a___},_,_,_,0,_,_,_,{1,___},_,_,_,_] := {RND,a};
   walk[{4,a___},_,_,_,0,_,_,_,{3,___},_,_,_,_] := {RND,a};
```

```
walk[{4,a___},_,_,_,0,_,_,_,_,_,_,_,{2,___}] := {RND,a};
walk[{4,a___},_,_,_,0,_,_,_,_,_,_,_,_] := 0;
walk[{_,a___},_,_,_,_,_,_,_,_,_,_,_,_] := {RND,a};
walk[0,{3,___},{4,___},_,_,_,_,_,_,_,_,_] := 0;
walk[0,{3,___},_,{1,___},_,_,_,_,_,_,_,_] := 0;
walk[0,{3,___},_,_,{2,___},_,_,_,_,_,_,_] := 0;
walk[0,_,{4,___},{1,___},_,_,_,_,_,_,_,_] := 0;
walk[0,_,{4,___},_,{2,___},_,_,_,_,_,_,_] := 0;
walk[0,_,_,{1,___},{2,___},_,_,_,_,_,_,_] := 0;
walk[0,{3,a___},_,_,_,_,_,_,_,_,_,_,_] := {RND,a};
walk[0,_,{4,a___},_,_,_,_,_,_,_,_,_,_] := {RND,a};
walk[0,_,_,{1,a___},_,_,_,_,_,_,_,_,_] := {RND,a};
walk[0,_,_,_,{2,a___},_,_,_,_,_,_,_,_] := {RND,a};
walk[0,_,_,_,_,_,_,_,_,_,_,_,_] := 0;
vonNeumann[func_, lat_] :=
  MapThread[func, Map[RotateRight[lat, #]&,
      {{0, 0}, {1, 0}, {0, -1}, {-1, 0}, {0, 1}}], 2];
GN[func_, lat_] :=
  MapThread[func, Map[RotateRight[lat, #]&,
        {{0, 0}, {1, 0}, {0, -1}, {-1, 0}, {0, 1},
         {1, -1}, {-1, -1}, {-1, 1}, {1, 1}, {2, 0},
         {0, -2}, {-2, 0}, {0, 2}}], 2];
NestList[GN[walk,
 Map[eval, vonNeumann[interact, #], {2}]]&, society, t]]
```

Group Interactions

JUDY: I know I'm different, but from now on I'm going to try and be the same.
HOWARD: The same as what?
JUDY: The same as people who aren't different.

—Barbra Streisand and Ryan O'Neal in *What's Up, Doc?* (1972)

5

Grouping and Conforming

"I hate it! I hate having to go along with everything my friends say."

—Molly Ringwald in *The Breakfast Club* (1985)

5.1 Introduction

Many social and economic situations can be viewed, as we have done in previous chapters, as bilateral interactions between two individuals. However, other social phenomena can be better described in terms of interactions between an individual and a group of other people. (We are not discussing so-called group-level interactions between clusters of people but rather interactions between an individual and two or more other individuals). In this chapter, we look at several phenomena that result from this sort of individual–group interaction, including neighborhood formation which can lead to voluntary segregation and local conformism which can result in the development of social norms and the spread of fads and fashions.

5.2 Forming Neighborhoods

A model for self-forming neighborhoods based on the desire of people to live with their *own kind* was proposed by Schelling [1978, 147–53] in one of the first applications of computer simulation in social science. The Schelling model, which allows an individual who is not happy with the number of nearest neighbors who are like him, to move to the nearest empty site that has a sufficient number of similar nearest neighbors, demonstrated that spatial segregation, or *ghettoization*, occurs spontaneously, without being imposed by a central authority. It can result in the clustering of people by gender or age at a social gathering or in the clustering of people by ethnicity or race in society at large.

Note: Because the Schelling model considers the movement of a person from where he is to be a result of his unhappiness with his neighbors, it is sometimes viewed as a model of an undesirable social phenomenon (fleeing). However, segregation, when it is voluntary, can have either positive or negative social and/or economic consequences [Durlauf 1997].

Here we recast the Schelling model so that an individual can decide to move, rather than stay where he is because of a desire to be with individuals who have similar attributes.

Note: We restrict movement to adjacent lattice sites here but in Chapter 6 we return to the Schelling model and allow nonlocal movements as well.

5.2.1 The System

Our model uses a square n by n lattice with wraparound boundary conditions. There is a population density p of individuals occupying lattice sites and the remaining sites are empty. The system evolves over a given number of time steps t.

5.2.2 Populating Society

The value of an empty site is 0.

The value of a site that is occupied by a person is an ordered pair. The first component of the ordered pair is a randomly chosen integer between 1 and 4 and indicates the direction the individual is facing (north, east, south, or west, respectively). The second component of the ordered pair is an *attribute list* of v elements, each of which has an integer value between 1 and w. The list elements represent various attributes of the individual and may include unchangeable traits such as race, gender, or ethnic identity and/or changeable beliefs such as political views, moral values, or personal interests.

Initially, the people are randomly placed on the lattice and face randomly chosen directions.

The code for this configuration is given by

```
RND := Random[Integer, {1, 4}]
society = Table[Floor[p + Random[]], {n}, {n}] /.
        1 :→ {RND, Table[Random[Integer, {1, w}], {v}]}
```

5.2.3 Executing a Time Step

During each time step, the following processes occur consecutively:

- each individual determines what fraction of his nearest neighbor sites are occupied by individuals whose attribute lists have at least 50% of their values in common with the individual and if the result is less than half, the individual decides to move; otherwise the individual decides to remain where he is; and

- each individual who has decided to move relocates to the adjacent nearest neighbor site he is facing, subject to the excluded volume constraint. All other individuals remain in place.

Deciding Whether to Stay or Move On

In this partial step, each individual decides whether he wants to stay in his present location or move. The criterion for this decision is based on the number of nearest neighbor sites containing individuals who share at least 50% of their attribute values with the individual.

For example, if the value of an individual is $\{2, \{2, 2, 2, 1, 2\}\}$ and the eight sites in its Moore neighborhood have attribute values $\{4, \{1, 2, 2, 1, 2\}\}$, $\{2, \{1, 1, 1, 2, 1\}\}$, 0, 0, $\{2, \{1, 1, 2, 2, 1\}\}$, $\{3, \{2, 2, 2, 2, 1\}\}$, 0, and $\{1, \{1, 1, 2, 1, 1\}\}$, then the individual shares 80%, 0%, 0%, 0%, 20%, 60%, 0%, and 40% of his attribute values with his nearest neighbor sites (an empty site has no attribute list and therefore shares no attribute values with a person) so that overall, the individual shares more than 50% of his attribute values with 25% of his eight nearest neighbor sites.

The movestay rule for an empty site is simply to remain where it is.

```
movestay[0, __] := 0
```

The left-hand side of the movestay for a site occupied by an individual can be written as

```
movestay[{a_, b_}, res__]
```

where b is the attribute list of the individual and res is the sequence consisting of the values of the eight nearest neighbor sites in the individual's Moore neighborhood.

For a site occupied by an individual, we must determine if a majority of the nearest neighbor sites share at least 50% of their attribute values with the person. This is done in several stages.

- The list of the number of shared attribute values between an individual and his north, east, south, west, northeast, southeast, southwest, and northwest neighbors, respectively, is determined using

```
Map[Count[b - #[[2]], 0]&, {res} /. 0 → {0, 0}]
```

Note: The value of an empty site is changed from 0 to {0, 0} in this calculation so that when its second component is used in (b - #[[2]]), a value of b is returned, and since b has no zero-valued elements, Count[b - #[[2]], 0]&[empty site] returns 0 indicating that the empty site has no shared attribute values with the person.

- The number of sites that have at least 50% of their attribute values in common with the person is determined using the Count function

```
Count[Map[Count[b - #[[2]], 0]&,
              {res} /. 0 → {0, 0}], _?(# >= v / 2 &)]
```

where _?(# >= v / 2 &) represents an expression (in this case, a number) whose value is equal to or greater than half the length v of the list.

The result of this calculation is a number that lies between 0, indicating that no neighbor sites share at least 50% of their attribute values with the individual, and 8, indicating that all of the neighbor sites share at least 50% of their attribute values with the individual.

Dividing this number by 8 and subtracting the result from 1 using

```
1 - Count[Map[Count[b - #[[2]], 0]&,
              {res} /. 0 → {0, 0}], _?(# >= v / 2 &)]/8.
```

gives the percentage of sites that share less than 50% of their attribute values with the individual. Applying the Round function to this,

```
Round[1 - Count[Map[Count[b - #[[2]], 0]&,
              {res} /. 0 → {0, 0}], _?(# >= v / 2 &)]/8.]
```

returns a value of 0 if more than 50% of the sites are occupied by individuals who share at least 50% of their attribute values with the individual and a value of 1 otherwise.

If we now take

```
{a* Round[1 - Count[Map[Count[b - #[[2]], 0]&,
              {res} /. 0 → {0, 0}], _?(# >= v / 2 &)]/8.], b}
```

we get {0, b} if the site is occupied by an individual with attribute list b who wants to stay where he is because at least 50% of his neighbor sites are occupied by individuals sharing at least 50% of their attribute values with the attribute values of the individual, or {a, b}, where a = 1, 2, 3, or 4 if the site is occupied by an individual who wants to move because not enough of his neighbors share enough of their attribute values with him.

We can therefore write the movestay rule for an individual (i.e., for a site occupied by an individual) as

```
movestay[{a_, b_}, res__] :=
  {a * Round[1 - Count[Map[Count[b - #[[2]], 0]&,
     {res} /. 0 → {0, 0}], _?(# >= v / 2 &)] / 8.], b}
```

The movestay rules are applied to the lattice sites using the anonymous function

```
Moore[movestay, #]&
```

where

```
Moore[func_, lat_] :=
  MapThread[func, Map[RotateRight[lat, #]&,
    {{0, 0}, {1, 0}, {0, -1}, {-1, 0}, {0, 1},
     {1, -1}, {-1, -1}, {-1, 1}, {1, 1}}], 2]
```

Moving

Once individuals have decided whether to stay where they are or move, the 28 rules for the movement of an individual are given by the set of walk rules in Chapter 1.

The walk rules are applied to the lattice sites using the anonymous function

```
GN[walk, Moore[movestay, #]]&
```

where

```
GN[func_, lat_] :=
  MapThread[func, Map[RotateRight[lat, #]&,
    {{0, 0}, {1, 0}, {0, -1}, {-1, 0}, {0, 1},
     {1, -1}, {-1, -1}, {-1, 1}, {1, 1}, {2, 0},
     {0, -2}, {-2, 0}, {0, 2}}], 2]
```

5.2.4 Evolving the System

The system evolves over t time steps, starting with the initial lattice configuration, society, using the following nesting operation.

```
NestList[GN[walk, Moore[movestay, #]]&, society, t]
```

5.2.5 The Program

n = size of lattice
p = population density
1, 2, ..., w = attribute values
v = number of values in attribute list
t = number of time steps

```
neighborhood[n_, p_, v_, w_, t_] :=
Module[{walk, movestay, society, RND, Moore, GN},

  RND := Random[Integer, {1, 4}];
  society = Table[Floor[p + Random[]], {n}, {n}] /.
          1 :> {RND, Table[Random[Integer, {1, w}], {v}]};

  movestay[0, __] := 0;
  movestay[{a_, b_}, res__] :=
    {a * Round[1 - Count[Map[Count[b - #[[2]], 0]&,
      {res} /. 0 -> {0, 0}], _?(# >= v / 2 &)] / 8.], b};

  (* walk rules go here *)

  Moore[func_, lat_] :=
    MapThread[func, Map[RotateRight[lat, #]&,
          {{0, 0}, {1, 0}, {0, -1}, {-1, 0}, {0, 1},
          {1, -1}, {-1, -1}, {-1, 1}, {1, 1}}]], 2];

  GN[func_, lat_] :=
    MapThread[func, Map[RotateRight[lat, #]&,
          {{0, 0}, {1, 0}, {0, -1}, {-1, 0}, {0, 1},
          {1, -1}, {-1, -1}, {-1, 1}, {1, 1}, {2, 0},
          {0, -2}, {-2, 0}, {0, 2}}]], 2];

  NestList[GN[walk, Moore[movestay, #]]&, society, t]]
```

5.2.6 Running the Simulation

We run the neighborhood program on a 20 by 20 lattice having a 60% population density of individuals, each of whom has one attribute that can have two possible values (thereby reducing the model to the Schelling model) over 500 time steps.

```
SeedRandom[9]
results = neighborhood[20, 0.60, 1, 2, 500];
```

We can compare the initial and final system configurations.

```
Show[GraphicsArray[
  Map[Show[Graphics[RasterArray[#  /.
    {0 -> RGBColor[0.7, 0.7, 0.7],
    {_, {1}} -> RGBColor[0, 1, 0],
    {_, {2}} -> RGBColor[0, 0, 1]}]],
    AspectRatio -> Automatic,
    DisplayFunction -> Identity]&,
  {First[results], Last[results]}]]];
```

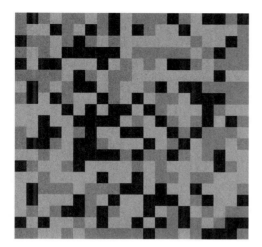 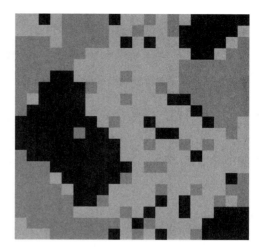

When there are three attributes, each having two possible values, the neighborhood picture is intriguing in terms of the tendency for certain colors, representing individuals with certain attribute list values, to appear next to one another.

 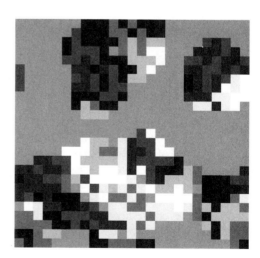

The preceding graphic was created with this code.

```
SeedRandom[37];
ghetto = neighborhood[25, 0.6, 3, 2, 500];
Show[GraphicsArray[
  Map[Show[Graphics[RasterArray[# /.
    {0 → RGBColor[0.7, 0.7, 0.7],
     {_, {1, 1, 1}} → RGBColor[1, 1, 1],
     {_, {1, 1, 2}} → RGBColor[1, 1, 0],
     {_, {1, 2, 1}} → RGBColor[1, 0, 1],
```

```
      {_, {2, 1, 1}} → RGBColor[0, 1, 1],
      {_, {2, 2, 1}} → RGBColor[0, 0, 1],
      {_, {2, 1, 2}} → RGBColor[0, 1, 0],
      {_, {1, 2, 2}} → RGBColor[1, 0, 0],
      {_, {2, 2, 2}} → RGBColor[0, 0, 0]}]],
    AspectRatio → Automatic,
    DisplayFunction → Identity]&,
  {First[ghetto], Last[ghetto]}]]];
```

Note: A similar graphics display can be created using the following code (which works for any values of *v* and *w*).

```
showCulture[lat_, v_, w_] :=
Module[{color},
  color[0] := Hue[0, 0, 1];
  color[{_, lis_}] := Hue[(1. + Apply[Plus,
    (lis - 1.) (w^Range[0, v - 1])]) / (w^v)];
  Show[Graphics[
    RasterArray[Map[color, lat, {2}]]],
    AspectRatio → Automatic]];

Scan[showCulture[#, 1, 2]&,
                {First[ghetto], Last[ghetto]}]
```

5.3 Conforming

In the neighborhood model, each individual uses his list of attribute values to decide whether to stay or move. An alternative to moving away from people who do not have a sufficient number of shared attribute values is to adopt the attribute values of these people. Although fixed traits such as race and gender can't be changed, beliefs, including beliefs about people with certain fixed traits, can be changed.

Here, we let each individual change his meme value(s) based on the meme values of all of the individuals around him. This models the process of conforming, in which an individual tries to *fit in* with the crowd. Conforming to a group and emulating a role model are both one-way value changes, however, they differ in that the latter involves interacting with a single individual and the former involves interacting with several other individuals. We look here at the change of meme values involving people who are spatial neighbors. In the next chapter, we examine meme changes involving people who form a *social neighborhood* consisting of individuals who either share some common meme values, such as friends, or who are related to one another, but are not necessarily adjacent to one another spatially.

On each time step in our model, every individual randomly chooses a meme value to change and then changes it to a value lying between its current value and the average value of that meme amongst its neighbors.

5.3.1 The System

Our model uses a square n by n lattice with wraparound boundary conditions. There is a population density p of individuals occupying lattice sites, and the remaining sites are empty. Each person is characterized by having a direction he is facing and a meme list with s elements. The system evolves over a given number of time steps t.

5.3.2 Populating Society

The value of an empty site is 0.

The value of a site occupied by an individual is an ordered pair, $\{x, y\}$ where:

- x is an integer value between 1 and 4, indicating the direction (N, E, S, and W) faced by the individual; and

- y is a list of s integer values, each ranging from 1 to m.

The initial system configuration consists of a randomly distributed population density p of individuals, each facing a randomly chosen direction. This configuration is created as follows.

```
RND := Random[Integer, {1, 4}]
society = Table[Floor[p + Random[]], {n}, {n}] /.
        1 :→ {RND, Table[Random[Integer, {1, m}], {s}]}
```

5.3.3 Executing a Time Step

Changing One's Meme Value

An individual picks a meme at random from his meme list and changes its value to a randomly chosen value between the current value and the average value of the meme amongst the people residing in the individual's Moore neighborhood.

The meme value that will be changed is chosen using

```
r = Random[Integer, {1, s}]
```

The average value of the rth meme amongst the neighbor sites is then calculated by first removing empty site values from the list of neighborhood site values *lis* using

```
DeleteCases[lis, 0]
```

Applying #[[2, r]]& to an individual's ordered pair value extracts its rth meme's value, and so we use

```
Map[#[[2, r]]&, DeleteCases[lis, 0]]
```

to create a list consisting of the *r*th meme values of all of the neighbors.

Finally, we calculate the average of these *r*th meme values using

```
(Apply[Plus, Map[#[[2, r]]&, #]]/Length[#])&[
                              DeleteCases[lis, 0]]
```

Overall, then, we can define a function that calculates the average value of the *r*th meme amongst the persons in the neighborhood of an individual.

```
memeAvg[lis_, r_] :=
   (Apply[Plus, Map[#[[2, r]]&, #]]/Length[#])&[
                              DeleteCases[lis, 0]]
```

The changing of the *r*th meme value in an individual's meme list *y* to an integer value randomly chosen between its current value and the neighborhood averaged value is accomplished using

```
ReplacePart[y,
         Random[Integer, {y[[r]], memeAvg[{res}, r]}], r]
```

Note: The memeAvg function can return a noninteger number, leading to pathological results in the case where there are only two possible meme values that differ by one.

Overall, we can write the meme value change rules as follows.

```
change[0, _, _, _, _, _, _, _, _] := 0
change[z_, 0, 0, 0, 0, 0, 0, 0, 0] := z
change[{a_, y_}, res__] :=
   Module[{r = Random[Integer, {1, s}]},
      {a, ReplacePart[y,
         Random[Integer, {y[[r]], memeAvg[{res}, r]}], r]}]
```

The first rule applies to an empty site that remains empty (with no meme list) regardless of the values of its neighbor sites. The second rule applies to an occupied site with no occupied neighbor sites (i.e., an isolated individual). The third rule applies to every site occupied by an individual and having at least one occupied neighbor site (i.e., to each individual having at least one neighbor).

The change rules are applied to the lattice sites using the anonymous function

```
Moore[change, #]&
```

where

```
Moore[func_, lat_] :=
   MapThread[func, Map[RotateRight[lat, #]&,
            {{0, 0}, {1, 0}, {0, -1}, {-1, 0}, {0, 1},
             {1, -1}, {-1, -1}, {-1, 1}, {1, 1}}], 2]
```

Deciding Whether to Move Or Stay

The rules for deciding whether to stay or attempt to move are given by the set of movestay rules shown in the previous model.

```
movestay[0, __] := 0
movestay[{a_, b_}, res__] :=
   {a * Round[1 - Count[Map[Count[b - #[[2]], 0]&,
         {res} /. 0 → {0, 0}], _?(# >= s / 2 &)] / 8.], b}
```

The movestay rules are applied to the lattice sites using the anonymous function

```
Moore[movestay, Moore[change, #]]&
```

Moving

The 28 rules for the movement of an individual are given by the set of walk rules in Chapter 1.

The walk rules are applied to the lattice sites using the anonymous function

```
GN[walk, Moore[movestay, Moore[change, #]]]&
```

where

```
GN[func_, lat_] :=
   MapThread[func, Map[RotateRight[lat, #]&,
         {{0, 0}, {1, 0}, {0, -1}, {-1, 0}, {0, 1},
          {1, -1}, {-1, -1}, {-1, 1}, {1, 1}, {2, 0},
          {0, -2}, {-2, 0}, {0, 2}}], 2]
```

5.3.4 Evolving the System

The system evolves over t time steps, starting with the initial lattice configuration, society, using the following nesting operation.

```
NestList[GN[walk,
         Moore[movestay, Moore[change, #]]]&, society, t]
```

5.3.5 The Program

n = size of lattice
p = population density
1, 2, ..., m = meme values
s = number of memes in meme list
t = number of time steps

```
conform[n_, p_, s_, m_, t_] :=
Module[{change, movestay, walk, society, RND, Moore, GN},

  RND := Random[Integer, {1, 4}];
  society = Table[Floor[p + Random[]], {n}, {n}] /.
       1 :> {RND, Table[Random[Integer, {1, m}], {s}]};
  memeAvg[lis_, r_] :=
    (Apply[Plus, Map[#[[2, r]]&, #]]/Length[#])&[
                            DeleteCases[lis, 0]];
  change[0, _, _, _, _, _, _, _, _] := 0;
  change[z_, 0, 0, 0, 0, 0, 0, 0, 0] := z;
  change[{a_, y_}, res__] :=
    Module[{r = Random[Integer, {1, s}]},
    {a, ReplacePart[y,
      Random[Integer, {y[[r]], memeAvg[{res}, r]}], r]}];

  movestay[0, __] := 0;
  movestay[{a_, b_}, res__] :=
    {a * Round[1 - Count[Map[Count[b - #[[2]], 0]&,
      {res} /. 0 -> {0, 0}], _?(# >= s / 2 &)] / 8.], b};

  (* walk rules go here *)

  Moore[func_, lat_] :=
    MapThread[func, Map[RotateRight[lat, #]&,
         {{0, 0}, {1, 0}, {0, -1}, {-1, 0}, {0, 1},
          {1, -1}, {-1, -1}, {-1, 1}, {1, 1}}], 2];
  GN[func_, lat_] :=
    MapThread[func, Map[RotateRight[lat, #]&,
         {{0, 0}, {1, 0}, {0, -1}, {-1, 0}, {0, 1},
          {1, -1}, {-1, -1}, {-1, 1}, {1, 1}, {2, 0},
          {0, -2}, {-2, 0}, {0, 2}}], 2];

  NestList[GN[walk,
       Moore[movestay, Moore[change, #]]]&, society, t]]
```

5.3.6 Running the Simulation

We run the conform program on a 20 by 20 lattice having a 60% population density
of individuals, each of whom has one attribute that can have two possible values over
500 time steps.

```
SeedRandom[9]
results = conform[20, 0.60, 1, 2, 500];
```

We can compare the initial and final system configurations.

```
Show[GraphicsArray[
  Map[Show[Graphics[RasterArray[# /.
    {0 -> RGBColor[0.7, 0.7, 0.7],
     {_, {1}} -> RGBColor[0, 1, 0],
     {_, {2}} -> RGBColor[0, 0, 1]}]],
    AspectRatio -> Automatic,
```

```
DisplayFunction → Identity]&,
{First[results], Last[results]}]]];
```

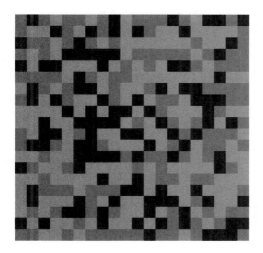

5.4 Social Norms, Fads, and Fashions

Social norms are commonly defined in terms of expectations, values, or behavior [Axelrod 1986]. The main characteristic of a social norm is that it involves agreement or consensus amongst a large number of individuals in a group without the intervention of a central authority.

Robert Axelrod [1986] has suggested several mechanisms that can give rise to and maintain a social norm, including membership which works by enabling like-minded individuals to interact with one another and excluding individuals who are not like-minded through a self-selection process (see Durlauf [1997] for some interesting ideas on membership and neighborhood). This process is evident in the preceding simulations.

5.5 Computer Simulation Projects

1. Schelling [1978] found that two modifications of his basic voluntary segregation model result in a close packing of the population of one type of person whereas the population of the other type of person is more spread out. The criteria were: (1) the requirement used by one type of person for being happy with his neighbors differs from the requirement used by another type of person for being happy with his neighbors, or (2) the population of one type of person is in the minority with respect to the population of the other type of person. Implement these two variations of the Schelling model into the neighborhood model and observe the behaviors of these variants.

2. Mobility is incorporated into the conforming model in this chapter. Modify the program so that individuals are immobile. Run the simulation and compare the result to that obtained when mobility is included. Look to see if there is a difference in the rate at which uniform values spread throughout the population.

3. Alter the conforming model so that the probability of an individual conforming is greater, the more the person differs from his neighbors. Observe the effect of this modification.

4. In the model of conforming in this chapter, each individual randomly chooses a meme value to bring into line with the average value of that meme amongst his neighbors. It might be more effective as a means of gaining acceptance to attempt to determine which meme is most important to the neighbors and then change the value of that meme.

Various criteria can be used to determine which meme is most important to the group. One indication of this might be that the more central a meme is to a group, the less its value will vary amongst the members. We can find the meme whose value varies least amongst the neighbors of an individual by summing the squares of the pairwise differences of all of the neighbors' values for each meme and taking the meme that returns the lowest value (e.g., if all of the neighbors agree on a particular meme, this sum will be zero). The computation for this is performed as follows.

We first form a list comprised of lists containing the values of the corresponding meme for each person in the neighborhood (e.g., the rth list will have all of the values of the rth meme in the neighborhood) using

```
Transpose[Transpose[DeleteCases[lis, 0]][[2]]]
```

We then determine the sum of the pairwise differences among the meme values in each of these lists using

```
Map[Apply[Plus, Flatten[Outer[Subtract, #, #]]^2]&,
          Transpose[Transpose[DeleteCases[lis, 0]][[2]]]]
```

We can illustrate how this operation works for a single list of meme values, $\{a, b, c\}$. (This might be the list of the values of the kth meme amongst the three people in the neighborhood around an individual; we use symbols rather than numerical values to make the explanation clearer.)

```
Flatten[Outer[Subtract, #, #]^2]&[{a, b, c}]
```

$\{0, (a-b)^2, (a-c)^2, (-a+b)^2, 0, (b-c)^2, (-a+c)^2, (-b+c)^2, 0\}$

```
Apply[Plus, Flatten[Outer[Subtract, #, #]^2]&[{a, b, c}]]
```

$(a-b)^2 + (-a+b)^2 + (a-c)^2 + (b-c)^2 + (-a+c)^2 + (-b+c)^2$

Note: This computation actually returns twice the sum of the *difference-squared* values which does not affect its usefulness for determining which meme has the smallest square distance value.

We can write these operations into a one-liner function definition

```
memeValueVariations[lis_] :=
  Map[Apply[Plus, Flatten[Outer[Subtract, #, #]]^2]&,
        Transpose[Transpose[DeleteCases[lis, 0]][[2]]]]
```

Finally, to determine which meme has the minimum difference-squared value, we take

```
Flatten[Position[#, Min[#]]]&[memeValueVariations[lis]]
```

which returns a list of the location(s) of the meme(s) that have the smallest difference-squared value (i.e., a return of {1, 3} indicates that the first and third memes both have the smallest difference-squared value). In the case where more than one meme has the same smallest difference-squared value, we randomly choose which meme to change by applying the anonymous function

```
#[[Random[Integer, {1, Length[#]}]]]&
```

to the list of locations so that overall we have

```
#[[Random[Integer, {1, Length[#]}]]]&[
  Flatten[Position[#, Min[#]]]&[memeValueVariations[lis]]]
```

Finally, we can incorporate this calculation into the conform program by replacing the third change rule

```
change[{a_, y_}, res__] :=
  Module[{r = Random[Integer, {1, s}]},
    {a, ReplacePart[y,
        Random[Integer, {y[[r]], memeAvg[{res}, r]}], r]}]
```

with the memeValueVariations definition and the change rule

```
memeValueVariations[lis_] :=
  Map[Apply[Plus, Flatten[Outer[Subtract, #, #]]^2]&,
        Transpose[Transpose[DeleteCases[lis, 0]][[2]]]]
change[{a_, y_}, res__] := Module[{r},
  r = #[[Random[Integer, {1, Length[#]}]]]&[
      Flatten[Position[#, Min[#]]]&[
      memeValueVariations[{res}]]];
    {a, ReplacePart[y,
        Random[Integer, {y[[r]], memeAvg[{res}, r]}], r]}]
```

Write the program for this model and compare the results it produces with the results obtained with the conform program.

5. The conform model combines the neighborhood model with a group average mechanism of individual meme change. One alternative to this is to use a pairwise meme exchange mechanism, as discussed in Chapter 2, with the neighborhood model. This can be done by incorporating the decide rules, exchange rules, and vonNeumann function from Chapter 2 into the neighborhood program, and by changing the first argument of the NestList operation to the following.

```
GN[walk, Moore[movestay,
          vonNeumann[exchange, vonNeumann[decide, #]]]]&
```

Carry out this modification, run the modified program, and compare the results obtained to the results of the conform model.

6. One model for the spread of ideas, fads, or fashions is suggested by the results of the role of social status described in Chapter 4: randomly assign social status values to individuals in the population. Randomly change one of the meme values of the individual in society with the highest social status. Then allow the new meme value to spread by changing, in each time step, the corresponding meme of the neighboring individual(s) with the highest social status to have the same value. Run this simulation to see the effect of the spread (or diffusion) of a meme value from the highest social status individuals to the individuals with the lowest social status.

7. Another model for the spread of fads and fashion based on social status would average over neighborhood meme values by weighting the meme value of a higher status individual more than the meme value of a lower status individual. Implement this idea into a program and see what behavior it shows over time.

8. Suppose that social status is endogenous, either increasing or decreasing as an individual becomes more like his neighbors. Run a simulation that incorporates this idea and see how the results differ for these two cases.

9. Axelrod [1986] has developed a behavioral-based model of social norms using the following definition: "a norm exists in a given social setting to the extent that individuals usually act in a certain way and are often punished when seen not to be acting in this way." In his model, individuals are either good or bad. If an individual is good, neither he nor anyone else in the society gets anything. If an individual is bad, he receives a positive payoff T (standing for the temptation to be bad) and all of the other individuals in the society receive a negative payoff H (standing for being hurt by the bad behavior). If an individual does behave badly, he is observed with a chance S (standing for being seen) by some of the other individuals. These individuals may impose a negative payoff P (standing for penalty) on the bad guy while they receive a negative payoff E (standing for the cost of enforcement) to themselves.

Each individual in the society has two attributes, boldness B and vengefulness V which are both real numbers between 0 and 1 [for computational purposes, each individual is assigned *B* and *V* values, each randomly chosen from eight possible values (0/7, 1/7, 2/7, 3/7, 4/7, 5/7, 6/7, 7/7)].

The following processes occur on each time step (with the resulting payoffs described previously).

- A value of *S* is chosen for each individual from a uniform distribution between 0 and 1, and an individual is bad whenever his *B* value is equal to or greater than *S* (i.e., whenever the chance of being seen by someone is less than the individual's boldness).

- Each individual sees a bad individual (except the bad guy himself) with probability *S*.

- Each individual who sees the defection punishes the defector with probability *V* (i.e., the more vengeful an individual is, the more likely he is to punish someone he sees being bad).

Implement this model.

10. The main shortcoming of the Axelrod model was anticipated by John Maynard Smith [1982] who concludes his book with the following statement (enclosed [...] words are added for clarification). "A group of individuals can agree not to [be bad] and to punish any member of the group who does [behave badly]. That, by itself, is not sufficient to guarantee [maintenance of the social norm] because the act of punishing is presumably costly and therefore individuals would be tempted to accept the benefits of [agreeing not to be bad] but not the costs of enforcing it [by punishing bad guys]." Smith offered a solution to this problem, suggesting that "[maintenance of the social norm] requires that refusal by an individual to participate in enforcing the [agreement] should also be regarded as a breach which will be punished."

This mechanism was adopted by Axelrod [1986] who refers to it as a *metanorm*. His metanorm model works as follows.

- A value of *S* is chosen for each individual from a uniform distribution between 0 and 1, and an individual behaves badly whenever his *B* value is equal to or greater than *S* (i.e., whenever the chance of being seen by someone is less than the individual's boldness).

- Each individual sees a person being bad (except the bad guy himself) with probability *S*.

- Each individual who sees the bad behavior punishes the bad guy with probability *V* (i.e., the more vengeful an individual is, the more likely he is to punish someone he sees being bad).

- Each individual sees a nonpunishing individual (except the nonpunisher) with probability *S*.

- Each individual who sees the nonpunishment punishes the nonpunisher with probability V.

Implement this model.

Note: The problem with the metanorm model of vengefulness against nonpunishers is that it often does not seem to work in the real world. For example, even the presence of a central authority to enforce the punishment of nonenforcers often fails to prevent the tolerance of bad behavior by others. This is evidenced by the frequent occurrence of cheating scandals in schools, including those that have an honor code in which not only cheaters but people who fail to report cheaters are punished. The prime reason for the failure of metanorms to work is that they fail to account for the penalty, often in the form of social ostracism, imposed on so-called squealers, within a group.

Note: Alternative means of maintaining a social norm, in addition to membership, have been suggested [Axelrod 1986; Tierney 1997], including the development of a social conscience [Frank 1987] and social ostracism. These mechanisms can be simulated using the ideas presented in Chapter 3.

5.6 References

Axelrod, Robert. 1986. "An Evolutionary Approach to Norms." *American Political Science Review* 80: 1095–1111.

Durlauf, Steven N. 1997. "The Memberships Theory of Inequality: Ideas and Implications." Sante Fe Institute Working Papers No. 97-05-047. Abstract.

Frank, Robert H. 1987. "If *Homo Economicus* Could Choose His Own Utility Function, Would He Want One with a Conscience?" *American Economic Review* 77: 593–604.

Schelling, Thomas C. 1978. *Micromotives and Macrobehavior*. New York: W. W. Norton.

Smith, John Maynard. 1982. *Evolution and The Theory of Games*. New York: Cambridge University Press, 173.

Tierney, John. 1997. "The Boor War." *The New York Times Magazine*, 5 January [includes statements attributed to Robert Axelrod].

5.7 Programs in the Chapter

5.7.1 neighborhood

```
neighborhood[n_, p_, v_, w_, t_] :=
Module[{walk, movestay, society, RND, Moore, GN},
  RND := Random[Integer, {1, 4}];
  society = Table[Floor[p + Random[]], {n}, {n}] /.
        1 :> {RND, Table[Random[Integer, {1, w}], {v}]}];
  movestay[0, __] := 0;
  movestay[{a_, b_}, res__] :=
    {a * Round[1 - Count[Map[Count[b - #[[2]], 0]&,
       {res} /. 0 -> {0, 0}], _?(# >= v / 2 &)] / 8.], b};
  walk[{1,a_},0,_,_,_,{4,_},_,_,_,_,_,_] := {RND,a};
  walk[{1,a_},0,_,_,_,_,_,{2,_},_,_,_,_] := {RND,a};
  walk[{1,a_},0,_,_,_,_,_,_,{3,_},_,_,_] := {RND,a};
  walk[{1,a_},0,_,_,_,_,_,_,_,_,_,_] := 0;
  walk[{2,a_},_,0,_,_,{3,_},_,_,_,_,_,_] := {RND,a};
  walk[{2,a_},_,0,_,_,_,{1,_},_,_,_,_,_] := {RND,a};
  walk[{2,a_},_,0,_,_,_,_,_,_,{4,_},_,_] := {RND,a};
  walk[{2,a_},_,0,_,_,_,_,_,_,_,_,_] := 0;
  walk[{3,a_},_,_,0,_,_,{4,_},_,_,_,_,_] := {RND,a};
  walk[{3,a_},_,_,0,_,_,_,{2,_},_,_,_,_] := {RND,a};
  walk[{3,a_},_,_,0,_,_,_,_,_,{1,_},_] := {RND,a};
  walk[{3,a_},_,_,0,_,_,_,_,_,_,_,_] := 0;
  walk[{4,a_},_,_,_,0,_,_,{1,_},_,_,_,_] := {RND,a};
  walk[{4,a_},_,_,_,0,_,_,_,{3,_},_,_,_] := {RND,a};
  walk[{4,a_},_,_,_,0,_,_,_,_,_,_,{2,_}] := {RND,a};
  walk[{4,a_},_,_,_,0,_,_,_,_,_,_,_] := 0;
  walk[{x_,a_},_,_,_,_,_,_,_,_,_,_,_] := {RND,a};
  walk[0,{3,_},{4,_},_,_,_,_,_,_,_,_,_] := 0;
  walk[0,{3,_},_,{1,_},_,_,_,_,_,_,_,_] := 0;
  walk[0,{3,_},_,_,{2,_},_,_,_,_,_,_,_] := 0;
  walk[0,_,{4,_},{1,_},_,_,_,_,_,_,_,_] := 0;
  walk[0,_,{4,_},_,{2,_},_,_,_,_,_,_,_] := 0;
  walk[0,_,_,{1,_},{2,_},_,_,_,_,_,_,_] := 0;
  walk[0,{3,a_},_,_,_,_,_,_,_,_,_,_] := {RND,a};
  walk[0,_,{4,a_},_,_,_,_,_,_,_,_,_] := {RND,a};
  walk[0,_,_,{1,a_},_,_,_,_,_,_,_,_] := {RND,a};
  walk[0,_,_,_,{2,a_},_,_,_,_,_,_,_] := {RND,a};
  walk[0,_,_,_,_,_,_,_,_,_,_,_] := 0;
  Moore[func_, lat_] :=
    MapThread[func, Map[RotateRight[lat, #]&,
          {{0, 0}, {1, 0}, {0, -1}, {-1, 0}, {0, 1},
           {1, -1}, {-1, -1}, {-1, 1}, {1, 1}}], 2];
  GN[func_, lat_] :=
    MapThread[func, Map[RotateRight[lat, #]&,
          {{0, 0}, {1, 0}, {0, -1}, {-1, 0}, {0, 1},
           {1, -1}, {-1, -1}, {-1, 1}, {1, 1}, {2, 0},
           {0, -2}, {-2, 0}, {0, 2}}], 2];
  NestList[GN[walk, Moore[movestay, #]]&, society,t]]
```

5.7.2 conform

```
conform[n_, p_, s_, m_, t_] :=
Module[{change, movestay, walk, society, RND, Moore, GN},
  RND := Random[Integer, {1, 4}];
  society = Table[Floor[p + Random[]], {n}, {n}] /.
         1 :> {RND, Table[Random[Integer, {1, m}], {s}]};
  memeAvg[lis_, r_] :=
    (Apply[Plus, Map[#[[2, r]]&, #]]/Length[#])&[
                                DeleteCases[lis, 0]];
  change[0, _, _, _, _, _, _, _, _] := 0;
  change[z_, 0, 0, 0, 0, 0, 0, 0, 0] := z;
  change[{a_, y_}, res__] :=
    Module[{r = Random[Integer, {1, s}]},
      {a, ReplacePart[y,
        Random[Integer, {y[[r]], memeAvg[{res}, r]}], r]}];
  movestay[0, __] := 0;
  movestay[{a_, b_}, res__] :=
    {a * Round[1 - Count[Map[Count[b - #[[2]]], 0]&,
      {res} /. 0 -> {0, 0}], _?(# >= s / 2 &)] / 8.], b};
  walk[{1,a_},0,_,_,_,{4,_},_,_,_,_,_,_] := {RND,a};
  walk[{1,a_},0,_,_,_,_,_,_,{2,_},_,_,_,_] := {RND,a};
  walk[{1,a_},0,_,_,_,_,_,_,_,{3,_},_,_,_] := {RND,a};
  walk[{1,a_},0,_,_,_,_,_,_,_,_,_,_,_] := 0;
  walk[{2,a_},_,0,_,_,_,{3,_},_,_,_,_,_,_] := {RND,a};
  walk[{2,a_},_,0,_,_,_,{1,_},_,_,_,_,_,_] := {RND,a};
  walk[{2,a_},_,0,_,_,_,_,_,_,_,{4,_},_,_] := {RND,a};
  walk[{2,a_},_,0,_,_,_,_,_,_,_,_,_,_] := 0;
  walk[{3,a_},_,_,0,_,_,_,{4,_},_,_,_,_,_] := {RND,a};
  walk[{3,a_},_,_,0,_,_,_,{2,_},_,_,_,_,_] := {RND,a};
  walk[{3,a_},_,_,0,_,_,_,_,_,_,{1,_},_,_] := {RND,a};
  walk[{3,a_},_,_,0,_,_,_,_,_,_,_,_,_] := 0;
  walk[{4,a_},_,_,_,0,_,_,_,{1,_},_,_,_,_] := {RND,a};
  walk[{4,a_},_,_,_,0,_,_,_,{3,_},_,_,_,_] := {RND,a};
  walk[{4,a_},_,_,_,0,_,_,_,_,_,_,_,{2,_}] := {RND,a};
  walk[{4,a_},_,_,_,0,_,_,_,_,_,_,_,_] := 0;
  walk[{x_,a_},_,_,_,_,_,_,_,_,_,_,_,_] := {RND,a};
  walk[0,{3,_},{4,_},_,_,_,_,_,_,_,_,_,_] := 0;
  walk[0,{3,_},_,{1,_},_,_,_,_,_,_,_,_,_] := 0;
  walk[0,{3,_},_,_,{2,_},_,_,_,_,_,_,_,_] := 0;
  walk[0,_,{4,_},{1,_},_,_,_,_,_,_,_,_,_] := 0;
  walk[0,_,{4,_},_,{2,_},_,_,_,_,_,_,_,_] := 0;
  walk[0,_,_,{1,_},{2,_},_,_,_,_,_,_,_,_] := 0;
  walk[0,{3,a_},_,_,_,_,_,_,_,_,_,_,_] := {RND,a};
  walk[0,_,{4,a_},_,_,_,_,_,_,_,_,_,_] := {RND,a};
  walk[0,_,_,{1,a_},_,_,_,_,_,_,_,_,_] := {RND,a};
  walk[0,_,_,_,{2,a_},_,_,_,_,_,_,_,_] := {RND,a};
  walk[0,_,_,_,_,_,_,_,_,_,_,_,_] := 0;
  Moore[func_, lat_] :=
    MapThread[func, Map[RotateRight[lat, #]&,
          {{0, 0}, {1, 0}, {0, -1}, {-1, 0}, {0, 1},
           {1, -1}, {-1, -1}, {-1, 1}, {1, 1}}], 2];
  GN[func_, lat_] :=
    MapThread[func, Map[RotateRight[lat, #]&,
```

```
          {{0, 0}, {1, 0}, {0, -1}, {-1, 0}, {0, 1},
           {1, -1}, {-1, -1}, {-1, 1}, {1, 1}, {2, 0},
           {0, -2}, {-2, 0}, {0, 2}}], 2];
NestList[GN[walk,
     Moore[movestay, Moore[change, #]]]&, society, t]]
```

Nonlocality

"No matter where you go, there you are."

—Peter Weller in *The Adventures of Buckaroo Banzai Across the 8th Dimension* (1984)

6

Social Networking and Moving to Far-Flung Locations

"Telephone call? All of these nuts could just make phone calls, they could spread insanity, oozing through telephone cables, oozing into the ears of all these poor sane people, infecting them. Wackos everywhere, plague of madness."

—Brad Pitt in *Twelve Monkeys* (1995)

6.1 Introduction

The use of spatial neighborhoods in modeling both the movement of individuals and their interactions is realistic in many situtations such as at a social gathering (a party) or in an organization (a company). However, it is too restrictive for modeling the sorts of economic and social phenomena that are increasingly taking place in the modern world where the tools of technology have fundamentally altered the nature of human interactions and movement. Today, a person can interact instantaneously with one or more people who are far away, using the telephone or email [a sister of one of the authors (RJG) first met her husband on the Internet] and the mass media can be used to propagate memes (the purpose of televangelism) or to spread gossip (talk shows create and maintain celebrity). People today can also move about from one location to another distant location (using an airplane or a car). In this chapter we present some basic tools that can be used for modeling nonlocal interactions and movement.

6.2 Reach Out and Touch Someone

In this model, which is a simple nonlocal version of the cultural transmission models in Chapter 2, each person randomly selects a friend (a member of his social network) to interact with and if the other person is agreeable, the two of them interact by changing the value of a randomly chosen meme.

6.2.1 The System

Our model uses an n by n square lattice, having a population density p of people occupying lattice sites; the remaining sites are empty. The system evolves over a given number of time steps t.

6.2.2 Populating Society

We first need to specify the values of the lattice sites:

- an empty site has value 0;

- a site that is occupied by a person has a value of a triplet of the form {name, socialLis, memeLis} in which the first component is a unique integer representing the person's name, the second component is a *social network* list consisting of the names of friends and/or relatives, and the third component is a *meme* list of s elements, each of which has an integer value between 1 and m.

The population is randomly distributed on the lattice, and initially the meme values of each person are randomly chosen.

This system is created in two steps, using

```
society = Table[Floor[p + Random[]], {n}, {n}] /.
  1 :> ++k /. x_?Positive :> {x, Union[DeleteCases[
            Table[Random[Integer, {1, k}], {f}], x]],
                Table[Random[Integer, {1, m}], {s}]};
society = society /. {a_, b_, c_} :> {a, Union[b,
      Cases[society, {x_, {___, a, ___}, _} :> x, {2}]], c}
```

The first part of the society definition creates a list of social network members for each person, independently. Thus, a person can have another individual in his social network without the other individual, in turn, having the person in his social network. Although this does happen, it seems more reasonable that, in general, any person who is in another individual's social network will have that individual in his social network. The second part of the society definition creates such a symmetric social network.

To illustrate the use of this operation, we consider a 4 by 4 lattice which is 25% occupied by individuals randomly placed on the lattice. Each agent has a meme list of two values, each of which is randomly chosen to be 1 or 2, and a list of friends and/or relatives randomly chosen from the population.

```
n = 4; p = 0.25; f = 3; m = 2; s = 2;
k = 0;
SeedRandom[51]
society = Table[Floor[p + Random[]], {n}, {n}] /.
   1 :> ++k /. x_?Positive :> {x, Union[DeleteCases[
                Table[Random[Integer, {1, k}], {f}], x]],
                Table[Random[Integer, {1, m}], {s}]};
society = society /. {a_, b_, c_} :> {a, Union[b,
      Cases[society, {x_, {___, a, ___}, _} :> x, {2}]], c}
```

$$
\begin{pmatrix}
\{1, \{3, 4\}, \{2, 1\}\} & 0 & 0 & 0 \\
\{2, \{3\}, \{2, 1\}\} & \{3, \{1, 2, 4\}, \{1, 1\}\} & 0 & 0 \\
\{4, \{1, 3\}, \{2, 2\}\} & 0 & 0 & 0 \\
0 & 0 & 0 & 0
\end{pmatrix}
$$

The four people in this example have the following symmetric social networks (notice that every person who appears on another person's social list, in turn, has that person on his social list):

- 1 has two friends and/or relatives $\{3, 4\}$,

- 2 has one friend and/or relative $\{3\}$,

- 3 has three friends and/or relatives $\{1, 2, 4\}$, and

- 4 has two friends and/or relatives $\{1, 3\}$.

6.2.3 Executing a Time Step

The time step is executed in two consecutive partial-steps. In the first partial-step, each person decides whom he wants to interact with in his social network, what meme will be altered, and what the meme's new value will be as a result of the interaction. In the second partial-step, cultural transmission takes place between persons who have mutually chosen to interact with each other. In both of these partial-steps, the update rules take as their argument a list consisting of the attribute lists of a person and of the people in the person's social network.

Note: We do not include any walk rules in this social network model because they are not needed if people are interacting with one another using some sort of telecommunication, such as email or telephone, rather than traveling to see them. Of course, movement can easily be incorporated into the model should that be desirable.

Deciding Which Friend to Interact with and What Cultural Change to Make

The following processes take place during the first partial-step.

- Each person randomly chooses an individual in his social network with whom to interact.

- Each person randomly chooses a meme.

- A new value for the chosen meme is calculated by randomly choosing an integer whose value lies between the value of the meme for the person and the value of the meme for the chosen individual.

The decide rule for persons who have a nonempty social network can be written as

```
decide[{{a_, b_, c_}, res__}] :=
  {a, b, c, {res}[[#2, 1]], #1,
    Random[Integer, {c[[#1]], {res}[[#2, 3, #1]]}]}]&[
               Random[Integer, {1, s}],
               Random[Integer, {1, Length[{res}]}]]
```

Note: In the anonymous function in the preceding decide rule, #1 is the position in the meme list that has been randomly chosen using Random[Integer, {1, s}] and #2 is the position of the individual in the person's social network who has been randomly chosen to interact with using Random[Integer, {1, Length[{res}]}].

The quantity {res}[[#2, 1]] determines the name of the chosen individual, #1 determines the meme that may be changed, and Random[Integer, {c[[#1]], {res}[[#2, 3, #1]]}] chooses a random integer between the value of the selected meme of the individual c[[#1]] and the value of the corresponding meme of the chosen person {res}[[#2, 3, #1]] to be the new value of the meme.

Applying the decide rule to a list consisting of a person and the members of his social network returns a six-tuple consisting of the following elements.

- The first element is the person's name.

- The second element is the person's social network list.

- The third element is the person's meme list.

- The fourth element is the name of the person who has been chosen for interaction.

- The fifth element is the location of the meme that will be affected by the interaction.

- The sixth element is the new value that the chosen meme will have as a result of the interaction.

Two additional decide rules are used to leave the values of empty sites and of loners who have no one in their social network unchanged.

```
decide[{{x_, {}, y_}}] := {x, {}, y}
decide[0] := 0
```

The decide rules are applied to the lattice sites using the anonymous function

```
socialCircle[decide, #]&
```

where

```
socialCircle[func_, lat_] :=
            Map[func[satellites[#, lat]]&, lat, {2}]
```

and

```
satellites[0, _] = 0
satellites[{a_, b_, c___}, lat_] := Join[{{a, b, c}},
        Cases[lat, {Apply[Alternatives, b], _, ___}, {2}]]
```

Transmitting Culture

The cultural exchange process is straightforward. If, in the decide partial-step, two people have chosen each other to interact with, the value of a corresponding meme changes for each person, according to the values *chosen by the person with the lower name value*. Otherwise no exchange occurs.

Note: Since the cultural exchange is based on the meme choices made by just one of two individuals who are interacting, it is somewhat inefficient to have each person choose a meme to change and a new value for that meme. However, it allows us to write simpler decide rules that actually execute rapidly.

The exchange rules for persons who have chosen another person who, in turn, has chosen them, can be written as

```
exchange[{{a_, b_, c_, aa_, pos_, d_},  ___,
        {aa_, _, _, a_, _, _},  ___}] :=
            {a, b, ReplacePart[c, d, pos]} /; a < aa
exchange[{{a_, b_, c_, aa_, _, _},  ___,
        {aa_, _, _, a_, pos_, d_},  ___}] :=
            {a, b, ReplacePart[c, d, pos]}
```

The exchange rules for loners and for empty sites are

```
exchange[{{a_, b_, c_, ___}, ___}] := {a, b, c}
exchange[0] := 0
```

The exchange rules are applied to the lattice sites using the anonymous function

```
socialCircle[exchange, socialCircle[decide, #]]&
```

6.2.4 Evolving the System

The system evolves over *t* time steps, starting with the initial lattice configuration, society, using the following nesting operation.

```
NestList[socialCircle[exchange,
                socialCircle[decide, #]]&, society, t]
```

6.2.5 The Program

n = size of lattice
p = population density
s = size of culture list (number of memes)
m = maximum cultural value (values from 1 to *m*)
f = maximum size of a social neighborhood (social circle)
t = number of time steps

```
peopleWhoNeedPeople[n_, p_, s_, m_, f_, t_] :=
Module[{k = 0, society, decide, exchange,
                        satellites, socialCircle},

  society = Table[Floor[p + Random[]], {n}, {n}] /.
    1 :> ++k /. x_?Positive :> {x, Union[DeleteCases[
            Table[Random[Integer, {1, k}], {f}], x]],
            Table[Random[Integer, {1, m}], {s}]};
  society = society /. {a_, b_, c_} :> {a, Union[b,
    Cases[society, {x_, {___, a, ___}, _} :> x, {2}]], c};

  decide[{{a_, b_, c_}, res__}] :=
    {a, b, c, {res}[[#2, 1]], #1,
      Random[Integer, {c[[#1]], {res}[[#2, 3, #1]]}]}&[
            Random[Integer, {1, s}],
            Random[Integer, {1, Length[{res}]}]]];
  decide[{{x_, {}, y_}}] := {x, {}, y};
  decide[0] := 0;

  exchange[{{a_, b_, c_, aa_, pos_, d_}, ___,
            {aa_, _, _, a_, _, _}, ___}] :=
            {a, b, ReplacePart[c, d, pos]} /; a < aa;
  exchange[{{a_, b_, c_, aa_, _, _}, ___,
            {aa_, _, _, a_, pos_, d_}, ___}] :=
            {a, b, ReplacePart[c, d, pos]};
  exchange[{{a_, b_, c_, ___}, ___}] := {a, b, c};
  exchange[0] := 0;

  socialCircle[func_, lat_] :=
            Map[func[satellites[#, lat]]&, lat, {2}];
  satellites[0, _] = 0;
  satellites[{a_, b_, c___}, lat_] := Join[{{a, b, c}},
      Cases[lat, {Apply[Alternatives, b], _, ___}, {2}]];
```

```
NestList[socialCircle[exchange,
              socialCircle[decide, #]]&, society, t]]
```

6.2.6 Running the Simulation

We run the program on a 20 by 20 lattice with a 70% population density for 500 time steps. Each person can have about five friends or relatives, as well as two memes, each of which has two possible values.

```
SeedRandom[31]
results = peopleWhoNeedPeople[20, 0.7, 2, 2, 5, 500];
```

The following graphic shows the population's meme content, sampled every 25 time steps. The different colors represent the four possible meme lists, {1, 1}, {1, 2}, {2, 1}, and {2, 2}.

```
<<Graphics`
counts[lat_] := Map[Count[lat, {_, _, _, #}, {2}]&,
              {{1, 1}, {1, 2}, {2, 1}, {2, 2}}];
Apply[PercentileBarChart,
  Transpose[Map[counts, results[[Range[1, 501, 25]]]]]];
```

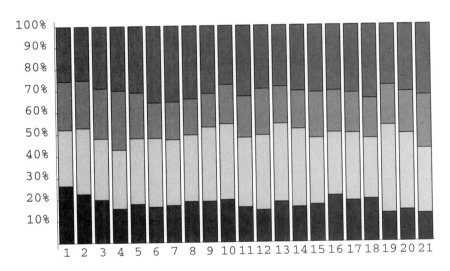

6.3 Follow the Leader

In this model, which is a nonlocal version of the conformism model in Chapter 5, each person changes the value of a randomly chosen meme to equal the value of the corresponding meme of the friend in his social network who has the highest social status.

6.3.1 The System

Our model uses an n by n square lattice, having a randomly distributed population density p of people occupying lattice sites; the remaining sites are empty. The system evolves over a given number of time steps t.

6.3.2 Populating Society

We first need to specify the values of the lattice sites:

- An empty site has value 0.

- A site that is occupied by a person has a value of a four-tuple of the form {name, socialLis, socialStatus, nemeLis} in which the first component is a unique integer representing the person's name, the second component is a *social network* list consisting of the names of up to f friends and/or relatives, the third component is the social status of the person as indicated by a real number between 0 and 1, and the fourth component is a *meme* list of s elements, each of which has an integer value between 1 and m.

This system is created using the following code (which differs from the code used to create society in the previous model only in including a social status value in the third position in a person's list).

```
society = Table[Floor[p + Random[]], {n}, {n}] /.
   1 :→ ++k /. x_?Positive :→ {x, Union[DeleteCases[
        Table[Random[Integer, {1, k}], {f}], x]],
          Random[], Table[Random[Integer, {1, m}], {s}]};
society = society /.
   {a_, b_, c_, d_} :→ {a, Union[b, Cases[society,
            {x_, {___, a, ___}, _, _} :→ x, {2}]], c, d}
```

6.3.3 Executing a Time Step

During a time step, each person randomly chooses one of his memes, locates the person in his social network who has the highest social status, and changes the value of the selected meme to equal the value of the corresponding meme of the friend with the highest social status.

This is accomplished using the following rules.

```
exchange[{{a_, b_, c_, d_}, res__}] := {a, b, c,
  ReplacePart[d,
        Sort[{res}, #1[[3]] > #2[[3]]&][[1, 4, #]], #]}&[
          Random[Integer, {1, s}]]
exchange[{{a_, {}, c_, d_}}] := {a, {}, c, d}
exchange[0] := 0
```

The second and third rules leave empty sites and loners (people with an empty social list) unchanged.

The first rule works as follows.

- Sort[{res}, #1[[3]] > #2[[3]]&] re-orders the list of the members of a person's social network in terms of decreasing social status.

- Sort[{res}, #1[[3]] > #2[[3]]&] [[1]] returns the person in the social network with the highest social status.

- Sort[{res}, #1[[3]] > #2[[3]]&] [[1, 4]] returns the meme list of the person in the social network with the highest social status.

- Sort[{res}, #1[[3]] > #2[[3]]&] [[1, 4, Random[Integer, {1, s}]]] returns the value of a randomly chosen meme for the person in the social network with the highest social status.

- ReplacePart[d, Sort[{res}, #1[[3]] > #2[[3]]&][[1, 4, #]], #]}&[Random[Integer, {1, s}]]] changes the value of the meme of an individual to the value of the corresponding meme for the person in the social network with the highest social status.

The exchange rules are applied to the lattice sites using the anonymous function

```
socialCircle[exchange, #]&
```

where

```
socialCircle[func_, lat_] :=
              Map[func[satellites[#, lat]]&, lat, {2}]
```

and

```
satellites[0, _] = 0
satellites[{a_, b_, c___}, lat_] := Join[{{a, b, c}},
      Cases[lat, {Apply[Alternatives, b], _, ___}, {2}]]
```

6.3.4 Evolving the System

The system evolves over t time steps, starting with the initial lattice configuration, society, using the following nesting operation.

```
NestList[socialCircle[exchange, #]&, society, t]
```

6.3.5 The Program

n = size of lattice
p = population density
s = size of culture list (number of memes)

m = maximum cultural value (values from 1 to m)
f = maximum size of a social neighborhood (social circle)
t = number of time steps

```
iWillFollowHim[n_, p_, s_, m_, f_, t_] :=
Module[{k = 0, society, exchange,
                        satellites, socialCircle},

   society = Table[Floor[p + Random[]], {n}, {n}] /.
     1 :> ++k /. x_?Positive :> {x, Union[DeleteCases[
         Table[Random[Integer, {1, k}], {f}], x]],
           Random[], Table[Random[Integer, {1, m}], {s}]};
   society = society /.
     {a_, b_, c_, d_} :> {a, Union[b, Cases[society,
         {x_, {___, a, ___}, _, _} :> x, {2}]], c, d};

   exchange[{{a_, b_, c_, d_}, res__}] := {a, b, c,
     ReplacePart[d,
         Sort[{res}, #1[[3]] > #2[[3]]&][[1, 4, #]], #]}&[
           Random[Integer, {1, s}]];
   exchange[{{a_, {}, c_, d_}}] := {a, {}, c, d};
   exchange[0] := 0;

   socialCircle[func_, lat_] :=
               Map[func[satellites[#, lat]]&, lat, {2}];
   satellites[0, _] = 0;
   satellites[{a_, b_, c___}, lat_] := Join[{{a, b, c}},
     Cases[lat, {Apply[Alternatives, b], _, ___}, {2}]];

   NestList[socialCircle[exchange, #]&, society, t]]
```

6.3.6 Running the Simulation

We run the program on a 20 by 20 lattice that is 70% full for 50 time steps. Each person can have about 15 friends, as well as two memes, each of which has two possible values.

```
SeedRandom[41]
results = iWillFollowHim[20, 0.7, 2, 2, 15, 50];

<<Graphics`
counts[lat_] := Map[Count[lat, {_, _, _, #}, {2}]&,
                     {{1, 1}, {1, 2}, {2, 1}, {2, 2}}];
Apply[PercentileBarChart, Transpose[Map[counts, results]]];
```

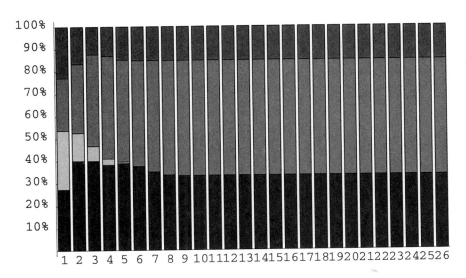

The graphic shows that the system becomes stable very quickly. After the first 10 time steps, no more meme exchange occurs. Thus, social norms are very quickly established in a small group whose members repeatedly interact with one another, such as a cult.

6.4 Nonlocal Movement

We illustrate the nonlocal movement of people using the variant of the Schelling model of self-forming neighborhoods that we developed in Chapter 5.

6.4.1 The System

Our model uses a square n by n lattice with wraparound boundary conditions. There is a population density p of individuals occupying lattice sites, and the remaining sites stay empty. The population consists of two types of individuals, of whom a fraction g are of one type and a fraction $(1 - g)$ are of the other type. The system evolves over a given number of time steps t.

6.4.2 Populating Society

The value of an empty site is 0.

The value of a site that is occupied by an individual is an attribute list consisting of one element whose value is 1 or 2, indicating what type of person he is.

Initially, the population is randomly distributed on the lattice.

The code for this is given by

```
society = Table[Floor[p + Random[]], {n}, {n}] /.
                    1 :> {1 + Floor[g + Random[]]}
```

6.4.3 Executing a Time Step

During each time step, the following processes occur consecutively.

- It is determined which individuals want to move from the sites they are occupying based on some criterion (the Schelling tipping model criterion for moving is that less than 50% of one's neighbors share the same attribute value) and to which empty sites they would be willing to move.

- Each individual who wants to move from the site he is on, and has a site he can move to, relocates to that location, leaving behind an empty site. All other sites remain unchanged.

Determining the Unhappy Campers and the Available Campsites

The first two rules change the value of a site that is occupied by an individual with the attribute list {1} to {1, 1} if that person wants to move, and likewise changes the value of a site that is occupied by an individual with the attribute list {2} to {2, 2} if that person wants to move, using for both types of people the criterion that he will want to move if he has fewer neighbors of his *own kind* than of the other kind.

```
decideToFly[{1}, res__] := {1, 1} /;
                Count[{res}, {1}] < Count[{res}, {2}]
decideToFly[{2}, res__] := {2, 2} /;
                Count[{res}, {2}] < Count[{res}, {1}]
```

The next two rules change the value of an empty site from 0 to {0, 1} or {0, 2} if it is an acceptable relocation site for a person of type {1} or {2}, respectively.

```
decideToFly[0, res__] := {0, 1} /;
                Count[{res}, {1}] > Count[{res}, {2}]
decideToFly[0, res__] := {0, 2} /;
                Count[{res}, {2}] > Count[{res}, {1}]
```

The last rule says that the value of all other occupied sites and empty sites remains unchanged.

```
decideToFly[x_, __] := x
```

The decideToFly rules are applied to the lattice sites using the anonymous function

```
Moore[decideToFly, #]&
```

where

```
Moore[func_, lat_] :=
  MapThread[func, Map[RotateRight[lat, #]&,
            {{0, 0}, {1, 0}, {0, -1}, {-1, 0}, {0, 1},
             {1, -1}, {-1, -1}, {-1, 1}, {1, 1}}], 2]
```

Moving

We first create a list, {*a1*, *a2*, *e1*, *e2*}, where

a1 — a list of the positions of all the unhappy people of type {1},

a2 — a list of the positions of all the unhappy people of type {2},

e1 — a list of the positions of empty sites that are suitable for a person of type {1}, and

e2 — a list of the positions of empty sites that are suitable for a person of type {2}.

This nested list is calculated using

```
{a1, a2, e1, e2} = Map[Position[lat, #, {2}]&,
                       {{1, 1}, {2, 2}, {0, 1}, {0, 2}}];
```

The number of unhappy campers who can move is restricted by the number of acceptable campsites that are available. To deal with this, we randomly match up unhappy campers with acceptable campsites until we run out of one or the other. Any remaining unhappy campers or acceptable campsites are left unchanged.

The following function can be used to randomize the order of the elements of a list.

```
randomize[{}] := {}
randomize[lis_] :=
  Transpose[Sort[Map[{Random[], #}&, lis]]][[2]]
```

Using the randomize function, we can take the lists *a1*, *a2*, *e1*, and *e2* and create four new lists, each having the same name as the original list from which it is constructed, and comprised of randomly chosen elements from the corresponding original list. The lengths of the new lists satisfy the condition that Length[*a1*] == Length[*e1*], and Length[*a2*] == Length[*e2*].

```
a1 = Take[randomize[a1], Min[Length[a1], Length[e1]]];
a2 = Take[randomize[a2], Min[Length[a2], Length[e2]]];
e1 = Take[e1, Length[a1]];
e2 = Take[e2, Length[a2]];
```

Each element in the new *a1* list is the position of a type {1} person who will move, and the corresponding element in the new *e1* list is the location to which that person will be moving. Similarly, each element in the new *a2* list is the position of a type {2} person who will move, and the corresponding element in the new *e2* list is the location to which that person will be moving.

To carry out the moves and leave behind empty sites, we define a function that replaces multiple parts of a lattice with values from a given list.

```
multiReplacePart[mat_, lis1_, lis2_] :=
  Fold[ReplacePart[#1, Part[#1, Apply[Sequence,
  First[#2]]], Last[#2]]&, mat, Transpose[{lis1, lis2}]]
```

Working with the entire lattice, we first occupy the empty sites listed in *e1* with people listed in *a1*, using

```
multiReplacePart[lat, a1, e1]
```

Having done this, we then occupy the empty sites listed in *e2* with people listed in *a2* using

```
multiReplacePart[#, a2, e2]&[
  multiReplacePart[lat, a1, e1]]
```

Finally, we convert the sites previously occupied by the people who have moved in the last two operations into empty sites having value 0, using

```
ReplacePart[#, 0, Join[a1, a2]]&[
    multiReplacePart[#, a2, e2]&[
      multiReplacePart[lat, a1, e1]]]
```

All that remains to be done is to eliminate the second component of the ordered pairs that were created when the decideToFly rules were applied to the lattice. This can be done using the replacement rules $\{0, \_\} :\to 0$ and $\{x\_, \_\} :\to \{x\}$.

Overall then, the moving rule is given by

```
flyAway[lat_] :=
Module[{a1, a2, e1, e2, randomize, multiReplacePart},

  {a1, a2, e1, e2} = Map[Position[lat, #, {2}]&,
                    {{1, 1}, {2, 2}, {0, 1}, {0, 2}}];
  randomize[{}] := {};
  randomize[lis_] :=
    Transpose[Sort[Map[{Random[], #}&, lis]]][[2]];
  a1 = Take[randomize[a1], Min[Length[a1], Length[e1]]];
  a2 = Take[randomize[a2], Min[Length[a2], Length[e2]]];
  e1 = Take[e1, Length[a1]];
  e2 = Take[e2, Length[a2]];

  multiReplacePart[mat_, lis1_, lis2_] :=
    Fold[ReplacePart[#1, Part[#1, Apply[Sequence,
    First[#2]]], Last[#2]]&, mat, Transpose[{lis1, lis2}]];

  ReplacePart[#, 0, Join[a1, a2]]&[
      multiReplacePart[#, a2, e2]&[
        multiReplacePart[lat, a1, e1]]] /.
                      {{0, _} :> 0, {x_, _} :> {x}}]
```

The flyAway rules are applied to the lattice sites using the anonymous function

```
flyAway[Moore[decideToFly, #]]&
```

6.4.4 Evolving the System

The system evolves over t time steps, starting with the initial lattice configuration, society, using the following nesting operation.

```
NestList[flyAway[Moore[decideToFly, #]]&, society, t]
```

6.4.5 The Program

n = size of lattice
p = population density
g = fraction of population with {1} value
t = number of time steps

```
flight[n_, p_, g_, t_] :=
Module[{society, decideToFly, flyAway, Moore},

  society = Table[Floor[p + Random[]], {n}, {n}] /.
                    1 :> {1 + Floor[g + Random[]]};

  decideToFly[{1}, res__] := {1, 1} /;
                Count[{res}, {1}] < Count[{res}, {2}];
  decideToFly[{2}, res__] := {2, 2} /;
                Count[{res}, {2}] < Count[{res}, {1}];
  decideToFly[0, res__] := {0, 1} /;
                Count[{res}, {1}] > Count[{res}, {2}];
  decideToFly[0, res__] := {0, 2} /;
                Count[{res}, {2}] > Count[{res}, {1}];
  decideToFly[x_, __] := x;

  flyAway[lat_] :=
  Module[{a1, a2, e1, e2, randomize, multiReplacePart},
   {a1, a2, e1, e2} = Map[Position[lat, #, {2}]&,
                          {{1, 1}, {2, 2}, {0, 1}, {0, 2}}];
   randomize[{}] := {};
   randomize[lis_] :=
     Transpose[Sort[Map[{Random[], #}&, lis]]][[2]];
   a1 = Take[randomize[a1], Min[Length[a1], Length[e1]]];
   a2 = Take[randomize[a2], Min[Length[a2], Length[e2]]];
   e1 = Take[e1, Length[a1]];
   e2 = Take[e2, Length[a2]];

   multiReplacePart[mat_, lis1_, lis2_] :=
    Fold[ReplacePart[#1, Part[#1, Apply[Sequence,
    First[#2]]], Last[#2]]&, mat, Transpose[{lis1, lis2}]];
```

```
      ReplacePart[#, 0, Join[a1, a2]]&[
        multiReplacePart[#, a2, e2]&[
          multiReplacePart[lat, a1, e1]]] /.
                          {{0, _} :→ 0, {x_, _} :→ {x}}];

  Moore[func_, lat_] :=
    MapThread[func, Map[RotateRight[lat, #]&,
          {{0, 0}, {1, 0}, {0, -1}, {-1, 0}, {0, 1},
           {1, -1}, {-1, -1}, {-1, 1}, {1, 1}}]], 2];

  NestList[flyAway[Moore[decideToFly, #]]&, society, t]]
```

6.4.6 Running the Simulation

We run the program for a 20 by 20 lattice with a 60% population density for 500 time
steps, using $v = 1$ (a single attribute) and $w = 2$ (two possible attribute values).

```
  SeedRandom[9];
  results = flight[20, 0.60, 0.5, 500];
```

We can compare the initial and final system configurations.

```
  Show[GraphicsArray[
    Map[Show[Graphics[RasterArray[# /.
        {0 → RGBColor[0.5, 0.5, 0.5],
         {1} → RGBColor[1, 0, 0],
         {2} → RGBColor[0, 0, 1]}]],
      AspectRatio → Automatic,
      DisplayFunction → Identity]&,
      {First[results], Last[results]}]]];
```

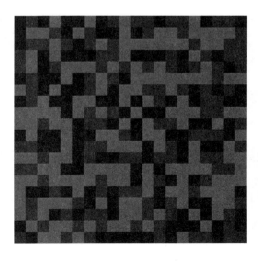 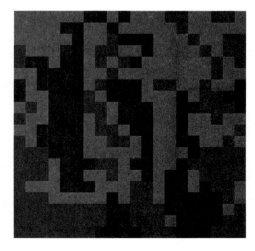

6.5 Computer Simulation Projects

1. Change the distribution of the sizes of social networks so that some individuals are especially charismatic, having many friends. How does the presence of these charismatic individuals affect the spread of new ideas?

2. Make the social status of an individual vary with the number of friends he has and see how this affects the results.

3. Let individuals have multiple (e.g., two) distinct sets of friends and let each operate on different memes. How do your findings change?

4. In the cultural transmission models, the social network remains the same over time. Modify the models so that people are added to or removed from the social network, based on a criterion of your choice.

6.6 References

Epstein, Joshua M. and Robert Axtell. 1996. *Growing Artifical Societies.* Cambridge MA: Brookings Institute Press/MIT Press.

6.7 Programs in the Chapter

6.7.1 peopleWhoNeedPeople

```
peopleWhoNeedPeople[n_, p_, s_, m_, f_, t_] :=
Module[{k = 0, society, decide, exchange,
                            satellites, socialCircle},
  society = Table[Floor[p + Random[]], {n}, {n}] /.
    1 :> ++k /. x_?Positive :> {x, Union[DeleteCases[
              Table[Random[Integer, {1, k}], {f}], x]],
                Table[Random[Integer, {1, m}], {s}]};
  society = society /. {a_, b_, c_} :> {a, Union[b,
    Cases[society, {x_, {___, a, ___}, _} :> x, {2}]], c};
  decide[{{a_, b_, c_}, res__}] :=
    {a, b, c, {res}[[#2, 1]], #1,
      Random[Integer, {c[[#1]], {res}[[#2, 3, #1]]}]}]&[
              Random[Integer, {1, s}],
              Random[Integer, {1, Length[{res}]}]]];
  decide[{{x_, {}, y_}}] := {x, {}, y};
  decide[0] := 0;
  exchange[{{a_, b_, c_, aa_, pos_, d_}, ___,
            {aa_, _, _, a_, _, _}, ___}] :=
              {a, b, ReplacePart[c, d, pos]} /; a < aa;
  exchange[{{a_, b_, c_, aa_, _, _}, ___,
            {aa_, _, _, a_, pos_, d_}, ___}] :=
              {a, b, ReplacePart[c, d, pos]};
  exchange[{{a_, b_, c_, ___}, ___}] := {a, b, c};
  exchange[0] := 0;
  socialCircle[func_, lat_] :=
              Map[func[satellites[#, lat]]&, lat, {2}];
  satellites[0, _] = 0;
  satellites[{a_, b_, c___}, lat_] := Join[{{a, b, c}},
    Cases[lat, {Apply[Alternatives, b], _, ___}, {2}]];
  NestList[socialCircle[exchange,
              socialCircle[decide, #]]&, society, t]]
```

6.7.2 iWillFollowHim

```
iWillFollowHim[n_, p_, s_, m_, f_, t_] :=
Module[{k = 0, society, exchange,
                            satellites, socialCircle},
  society = Table[Floor[p + Random[]], {n}, {n}] /.
    1 :> ++k /. x_?Positive :> {x, Union[DeleteCases[
        Table[Random[Integer, {1, k}], {f}], x]],
          Random[], Table[Random[Integer, {1, m}], {s}]};
  society = society /.
    {a_, b_, c_, d_} :> {a, Union[b, Cases[society,
          {x_, {___, a, ___}, _, _} :> x, {2}]], c, d};
  exchange[{{a_, b_, c_, d_}, res__}] := {a, b, c,
    ReplacePart[d,
        Sort[{res}, #1[[3]] > #2[[3]]&][[1, 4, #]], #]}&[
```

```
                        Random[Integer, {1, s}]];
    exchange[{{a_,   {}, c_, d_}}] := {a, {}, c, d};
    exchange[0] := 0;
    socialCircle[func_, lat_] :=
                Map[func[satellites[#, lat]]&, lat, {2}];
    satellites[0, _] = 0;
    satellites[{a_, b_, c___}, lat_] := Join[{{a, b, c}},
        Cases[lat, {Apply[Alternatives, b], _, ___}, {2}]];
    NestList[socialCircle[exchange, #]&, society, t]]
```

6.7.3 flight

```
flight[n_, p_, g_, t_] :=
Module[{society, decideToFly, flyAway, Moore},
    society = Table[Floor[p + Random[]], {n}, {n}] /.
                            1 :> {1 + Floor[g + Random[]]};
    decideToFly[{1}, res__] := {1, 1} /;
                    Count[{res}, {1}] < Count[{res}, {2}];
    decideToFly[{2}, res__] := {2, 2} /;
                    Count[{res}, {2}] < Count[{res}, {1}];
    decideToFly[0, res__] := {0, 1} /;
                    Count[{res}, {1}] > Count[{res}, {2}];
    decideToFly[0, res__] := {0, 2} /;
                    Count[{res}, {2}] > Count[{res}, {1}];
    decideToFly[x_, __] := x;
    flyAway[lat_] :=
    Module[{a1, a2, e1, e2, randomize, multiReplacePart},
    {a1, a2, e1, e2} = Map[Position[lat, #, {2}]&,
                        {{1, 1}, {2, 2}, {0, 1}, {0, 2}}];
    randomize[{}] := {};
    randomize[lis_] :=
      Transpose[Sort[Map[{Random[], #}&, lis]]][[2]];
    a1 = Take[randomize[a1], Min[Length[a1], Length[e1]]];
    a2 = Take[randomize[a2], Min[Length[a2], Length[e2]]];
    e1 = Take[e1, Length[a1]];
    e2 = Take[e2, Length[a2]];
    multiReplacePart[mat_, lis1_, lis2_] :=
     Fold[ReplacePart[#1, Part[#1, Apply[Sequence,
     First[#2]]], Last[#2]]&, mat, Transpose[{lis1, lis2}]];
    ReplacePart[#, 0, Join[a1, a2]]&[
        multiReplacePart[#, a2, e2]&[
        multiReplacePart[lat, a1, e1]]] /.
                        {{0, _} :> 0, {x_, _} :> {x}}];
    Moore[func_, lat_] :=
      MapThread[func, Map[RotateRight[lat, #]&,
            {{0, 0}, {1, 0}, {0, -1}, {-1, 0}, {0, 1},
            {1, -1}, {-1, -1}, {-1, 1}, {1, 1}}], 2];
    NestList[flyAway[Moore[decideToFly, #]]&, society, t]]
```

Appendices

"Either you go by the rules or you lose."

—Walter Matthau in *Lonely Are the Brave* (1962)

Appendix A

How *Mathematica* Works

In this appendix, we describe some of the underpinnings of the *Mathematica* programming language. Understanding how the *Mathematica* language works will enable you to better understand the programs presented in this book (illustrations of the use of all of the *Mathematica* built-in functions that are employed in this book are given in Appendix B). This is not meant as a complete overview of *Mathematica*. This material has been extensively *field tested* in courses presented in universities, professional conferences, and industry.

A.1 Expressions

A.1.1 Everything Is an Expression

Underlying every *Mathematica* command is a *Mathematica* expression. There are two parts to every *Mathematica* expression: the *head* and the *arguments*. The syntax for an expression is to follow the head with the arguments surrounded with square brackets.

```
head[ arg1, arg2, ....., argn ]
```

The head is itself a *Mathematica* expression, and the arguments consist of a sequence of zero or more expressions separated by commas. For certain heads, there are more familiar shorthands which may be used but, internally, *Mathematica* keeps track of the head and arguments as if they were in the preceding form. For example, when adding number or symbols together you can use the familiar *infix* notation

```
a + b + c
```

However, when you enter this, it is transformed by *Mathematica* into an expression with Plus as the head and the objects that are being added as arguments. One way to see this is by asking *Mathematica* for the full internal form of the expression

```
FullForm[a + b + c]
```

Plus[a, b, c]

A.1.2 Noteworthy Expressions

Our previous description of expressions as anything of the form *head[arguments]* where head and arguments are themselves expressions may seem circular because it is not complete. What's missing is a discussion of some special types of expressions that play an essential role in the evaluation of expressions and in the writing of *Mathematica* programs.

Atoms

An atomic expression is an expression that cannot be broken down further into other expressions. There are several kinds of atomic expressions

- A *number*: the four types of numbers that *Mathematica* handles by default are integers, real numbers, complex numbers, and rational numbers.

- A *string*: any sequence of characters enclosed in quotes, as in "Simulating Society".

- A *symbol*: a letter followed by letters and/or numbers that are not enclosed in quotes. We show shortly how symbols can be used to store numbers, strings, or arbitrary expressions.

When you ask for the FullForm of an atomic expression, you will notice that the expression itself is returned.

```
{FullForm[Plus],
        FullForm["Simulating Society"], FullForm[51]}
```

{Plus, "Simulating Society", 51}

Furthermore, the head of an atomic expression describes what kind of atom it is.

```
{Head[Plus], Head["Simulating Society"], Head[51]}
```

{Symbol, String, Integer}

CompoundExpression

Another very important kind of expression is a CompoundExpression. A program written in any computer language usually consists not of a single command, but of many commands one following another. When the program is run, these commands are typically executed consecutively in the order in which they are written, and the final result of these executions is displayed on the computer screen, or perhaps sent to a printer. In *Mathematica*, this is done with a CompoundExpression. A CompoundExpression is a series of expressions separated by semicolons. When a CompoundExpression is evaluated, all of the expressions within it are evaluated successively, and the output from the last evaluation is returned (unless the last expression is followed by a semicolon, in which case nothing is returned). This is how all the programs in this book are constructed — as CompoundExpressions.

In this example, all the expressions are evaluated in turn (the symbol a is assigned a value equal to the sum of 5 and 8 and is then multiplied by 7) but only the result of evaluating the final expression after which there is no semicolon returns visible output.

```
a = 5 + 8; 7 a
```

91

A.2 Patterns

When working with expressions, it is often useful to be able to determine the general form or structure of an expression without worrying about the details of its content. For instance, pattern-matching is used to cause a function to perform one action when applied to a quantity having a certain form, say a number, and another action when applied to a quantity having a different form, say a symbol. In fact, all of the functions used in *Mathematica,* both those that are a part of *Mathematica* itself, known as *built-in* functions, and the *user-defined* functions that we create, which includes the programs we write, deal with patterns on some level, as pattern-matching is central to how evaluation takes place. We cover more of the details of *Mathematica* evaluation later in this appendix.

To give a real-world analogue of pattern-matching in the English language, there are certain patterns of words that make up sentences. One such basic pattern might be *noun-verb-noun*, representing any sentence that consists of a noun followed by a verb followed by a noun. A sentence that matches this grammatical pattern is "Suzanne reads books." A sentence that does not match this pattern is "Carole writes." So too in *Mathematica*, a *pattern* is an expression that represents all expressions that share a particular syntactic structure. The syntactic structure of an expression is given by the FullForm representation of the expression, which we discussed earlier.

There are a number of *Mathematica* functions that allow one to use patterns. One such important function is MatchQ, which determines whether a given expression matches a pattern.

```
?MatchQ
```

MatchQ[expr, form] returns True if the pattern form
matches expr, and returns False otherwise.

A.2.1 Matching Expressions

One of the simplest patterns matches any expression, and is represented in *Mathematica* by a single underscore character (_). This pattern is known as a Blank. Since a Blank matches any expression, if MatchQ is given any expression as its first argument and a Blank as its second argument, it will always return True.

```
MatchQ[a + b, _]
```

True

One can also use a Blank to match expressions that lie within other expressions. Such expressions are called *subexpressions*. For example, the following input returns True because the first Blank matches the first 1, and the last Blank matches the 3. The expressions 1 and 3 are simple subexpressions of the given expression.

```
MatchQ[func[1, 1, 1, 2, 2, 3], func[_, 1, 1, 2, 2, _]]
```

True

Note: There is an exception to the matching of Blank that is important to keep in mind. Although each of the individual subexpressions that reside within the sequence subexpression used in the previous example matches Blank, the sequence itself does not match Blank.

```
MatchQ[func[1, 1, 1, 2, 2, 3], func[_]]
```

False

One can restrict the Blank to match only expressions with a specified head by giving that head to Blank as an argument. The shorthand form of Blank[*h*] is an underscore followed by the symbol *h*, as in _*h*. For example, the pattern in the following matches any expression with head *func* and an integer argument.

```
MatchQ[func[36], func[_Integer]]
```

True

Another modification of Blank is to assign a name to a pattern. You can name a pattern by inserting a symbol before the underscore, with no intervening space. As a simple example, the pattern $x\_$ matches any expression and is named with the dummy variable x. Providing the pattern with a name makes it easy to refer to the part of an expression that matches a part of a pattern. These sorts of references are indispensable for defining your own functions, where the names are used as *dummy variables*. We explain function writing later, but for purposes of illustrating the use of pattern labeling, the simple function definition $g[x\_] := a + x$ can be used. If we first create the function definition using the labeled pattern $x\_$ on the left-hand side and then enter $g[7]$, the value of 7 is substituted for the symbol x on the right-hand side of the function definition and the value $7 + a$ is returned.

```
g[x_] := a + x; g[7]

7 + a
```

A.2.2 Matching Sequences of Expressions

There are two extensions of Blank that are widely used: BlankSequence and BlankNullSequence.

BlankSequence is represented by two underscore characters in a row (__), and matches any sequence of one or more comma-separated *Mathematica* expressions. To illustrate the use of BlankSequence, we can use func[1, 1, 1, 2, 2, 3] which we saw earlier does not match func[_].

```
{MatchQ[func[1, 1, 1, 2, 2, 3], func[_]],
 MatchQ[func[1, 1, 1, 2, 2, 3], func[__]]}

{False, True}
```

BlankNullSequence, represented by three underscore characters (___), matches any sequence of *zero* or more comma-separated *Mathematica* expressions. To illustrate the use of BlankNullSequence, we can use func with no arguments.

```
{MatchQ[func[], func[_]],
 MatchQ[func[], func[__]],
 MatchQ[func[], func[___]]}

{False, False, True}
```

Blank, BlankSequence, and BlankNullSequence are in order from most specific to most general. That is, every expression that matches _ will also match __ and ___. Furthermore, every expression that is a sequence of one or more expressions (and so matches __) is also a sequence of zero or more expressions (and so will match ___). As an example, we can look at the expression $a + b$ which we saw earlier matches Blank.

```
{MatchQ[a + b, _], MatchQ[a + b, __], MatchQ[a + b, ___]}

{True, True, True}
```

Choosing which of these patterns to use allows us to make fine distinctions between expressions which is very useful in creating function definitions that have the same name but give different results when applied to different arguments.

Here is an example where the __ matches a sequence of five expressions, and the ___ matches a sequence of zero expressions.

```
MatchQ[func[1, 1, 1, 2, 2, 3], func[__, 3, ___]]
```

```
True
```

If we switch the use of BlankSequence with BlankNullSequence, the match fails because there are no arguments after the 3 in the given expression.

```
MatchQ[func[1, 1, 1, 2, 2, 3], func[___, 3, __]]
```

```
False
```

This pattern will match any application of the function func whose sequence of arguments begins with 1, contains 2, and ends with 3.

```
MatchQ[func[1, 1, 1, 2, 2, 3], func[1, ___, 2, ___, 3]]
```

```
True
```

As with Blank, the patterns BlankSequence and BlankNullSequence can be restricted to match sequences of expressions each of which has the same head.

```
MatchQ[func[1, 1, 1, 2, 2, 3], func[exprs___Integer]]
```

```
True
```

A.2.3 Alternatives

Alternatives[*a1*, *a2*, ...] is a pattern that matches an expression when any one of the patterns *a1*, *a2* ... matches it. The shorthand for Alternatives is a vertical bar (|), so that Alternatives[*a*, *b*, *c*] can be rewritten *a* | *b* | *c*. The Alternatives function can be thought of as a *multiple-choice* pattern. Using Alternatives, we can provide a pattern that matches a more general set of expressions than each of the patterns alone.

The pattern in the following example would match any expression with head func that has one or more arguments, and whose last argument was one of 1, 3, or 5.

```
MatchQ[func[1, 1, 1, 2, 2, 3], func[__, 1 | 3 | 5]]
```

```
True
```

Continuing with the analogy to multiple-choice, an expression that returns False when compared to the Alternatives pattern is equivalent to the multiple-choice option *none of the above*.

```
MatchQ[func[1, 1, 1, 2, 2, 3], func[___, 1 | 2 | 5]]
```

False

A.2.4 PatternTest

In addition to restricting a pattern so that it only matches expressions with a certain head, there are several other ways that *Mathematica* can restrict the set of expressions that match a given pattern. One simple way is through the PatternTest function. PatternTest[_, *func*] matches an expression *expr* when *func*[*expr*] returns True. Functions that return True or False when applied to an expression are commonly called predicate functions. A shorthand for PatternTest is to follow an underscore (or two underscores or three underscores) with a question mark, and the name of the predicate function as in _?*func*. For example, the *Mathematica* function Positive returns True when applied to a positive real number.

```
MatchQ[3, _?Positive]
```

True

We can combine the use of the head restriction with the use of PatternTest to narrow the pattern matching even further since if either of these restrictions is not met for a given expression, the pattern will fail to match. In general, the argument of Blank (or BlankSequence or BlankNullSequence) can be used with PatternTest to create a new pattern of the form PatternTest[_*head*, *func*] or as it is commonly written _*head*?*func*. For example, using _Integer which matches integers, with ?Positive which matches positive numbers creates a pattern that matches positive integers.

```
{MatchQ[2.71, _Integer],
 MatchQ[2.71, _?Positive],
 MatchQ[2.71, _Integer?Positive]}
```

{False, True, False}

A.2.5 Condition

The Pattern test function is applied to an entire expression. If we only want to pattern-match a part of an expression, we use the Condition function.

Condition takes two arguments: the pattern and the condition. If the condition returns True, then we say that the condition is satisfied, and the pattern-match is successful. Condition[*p*, *cond*] is usually written as *p* /; *cond*. For example, the following returns

True because the first argument to *func* is a positive number that is less than the last argument.

```
MatchQ[func[1, 1, 1, 2, 2, 3],
               func[x_?Positive, ___, z_] /; x < z]

True
```

As this example shows, Condition contains labeled patterns in order to identify the part of the pattern that is being used in the pattern-matching restriction.

A.2.6 Rule, RuleDelayed, and ReplaceAll

Patterns can be used not only to determine if an expression is of a specified form, but also to transform the matching subexpressions into other expressions. In order to do this, we can specify a Rule called a *replacement rule* or *transformation rule*, Rule[*p*, *q*], which specifies a pattern *p* and a value *q* with which to replace the pattern. This is commonly written as $p \to q$. Another function, ReplaceAll, is usually used to apply the Rule function to an expression and carry out the replacement in the expression. ReplaceAll[*expr*, Rule[*p*, *q*]] replaces *p* in any subexpression of *expr* with *q*. This is usually written with the shorthand *expr* /. $p \to q$. Here is an example.

```
func[-3, -2, -1, 0, 1, 2, 3] /. x_?Negative → x^2

func[9, 4, 1, 0, 1, 2, 3]
```

Notice that without naming the pattern on the left-hand side of the replacement rule, there would be no way to refer to it on the right-hand side. If you also wanted to replace the positive numbers by their cubes, you could do it by specifying a list of replacement rules.

```
func[-3, -2, -1, 0, 1, 2, 3] /.
               {x_?Negative → x^2, x_?Positive → x^3}

func[9, 4, 1, 0, 1, 8, 27]
```

A technical note about the evaluation of ReplaceAll: when an expression of the form *expr* /. $p \to q$ is entered, first *expr* is evaluated, then *p* is evaluated, then *q* is evaluated, and finally the replacement is carried out. Although this is fine for many instances, there may be cases where you want the evaluation of *q* to wait until after the replacement is actually carried out (in other words, so that the evaluated *p* pattern in the evaluated *expr* is replaced with the unevaluated *q* value). The built-in function RuleDelayed does exactly this. Instead of an arrow, RuleDelayed is entered using the "colon-arrow" :→ shorthand.

To illustrate this, we use the built-in Random function which returns a random real number between zero and one. If we use a replacement rule that evaluates the random number first, we see that every integer in the following expression is replaced with the same number.

```
func[1, 2, 3] /. x_Integer → Random[]

func[0.810421, 0.810421, 0.810421]
```

However, using RuleDelayed (:→), postpones the call to Random until the rule is used to replace each integer individually. Thus, the result contains three distinct random numbers.

```
func[1, 2, 3] /. x_Integer :→ Random[]

func[0.174746, 0.49314, 0.234172]
```

Finally, we can use Condition in conjunction with RuleDelayed (but not with Rule) to provide patterns that will only be used as replacement rules if some predicate is satisfied. For example, the following is a simple method to replace all integers with absolute value greater than 1 with their squares

```
func[-3, -2, -1, 0, 1, 2, 3] /.
                    x_Integer :→ x^2 /; Abs[x] > 1

func[9, 4, -1, 0, 1, 4, 9]
```

A.3 Functions

A.3.1 Built-In Functions

Before defining your own functions, it is a good idea to become familiar with what sorts of things *Mathematica* can already do with the functions that are built into the *Mathematica* kernel, each of which has a particular syntax and usage. Although it is completely unnecessary to become familiar with all 1,500+ of them, writing your own simulations will be easier if you make an effort to understand all the built-in functions used in this book. You can find out more about a built-in function by asking *Mathematica* for its *usage message*, which you can do by typing a question mark (?) and then the function name. For example,

```
?Positive

Positive[x] gives True if x is a positive number.
```

You can get more information by looking up these functions in the *Mathematica* book, or in the Built-in Functions section of the Help Browser in the *Mathematica* front end.

Many built-in functions have special formats that make entering and reading them easier. A list of such special formats is available in the *Mathematica* book.

A.3.2 User-Defined Functions

Using a *function* or *program* in *Mathematica* is simple, as we've already seen: you input the function name and a sequence of arguments, and the kernel returns a result (the details of how *Mathematica* evaluates the input are discussed later in this appendix). So to define a new function in *Mathematica*, you need to specify the name of the function (a symbol), the arguments to the function (a pattern or sequence of patterns, usually labeled), and then what the function does to those arguments. Here is an example of a function that squares its argument, and then adds one to the result.

```
func1[x_] := x^2 + 1
```

Note: You may have noticed that built-in functions begin with a capital letter, and capitalize every complete word thereafter. To distinguish user-defined functions from built-in functions, we follow the convention of beginning our function definitions with a lowercase letter.

The "colon-equal" (:=) in the function definition looks very much like the "colon-arrow" (:→) from the previous section, in that we have a pattern on the left-hand side that is transformed to a different expression on the right-hand side. The difference between using := (or SetDelayed) and :→ (RuleDelayed) is in how the replacement is applied. The replacement that occurs with RuleDelayed is restricted to the expression to which it is attached. With SetDelayed, the replacement occurs automatically for every expression you input that pattern-matches the left-hand side of the definition.

```
func1[5]
```

> *26*

Functions can be applied one right after another without having to return intermediate results.

```
func1[func1[5]]
```

> *677*

Of course, unlike the built-in functions, user-defined functions do not persist once you quit *Mathematica,* so if you want to use the function again in another *Mathematica* session, you must first re-enter the definition.

Along with function definitions, it is sometimes useful to assign your own values to a symbol, so that the symbol becomes a shorthand for the value. This can be done using the = (Set) operator. For example, after making the following assignment, *b* will be replaced by 23 wherever it appears. (Note the use of a semicolon, the CompoundExpression operator, to prevent the output from being given.)

```
b = 23;
```

Now this example will replace b with its assigned value as it is evaluated.

```
func1[a] + b
```

$24 + a^2$

A program is nothing more than a user-defined function. For example, *func1* is a program. Of course, a program often looks complicated because the right-hand side of a program is usually a CompoundExpression. As an example, we can write a simple program that allows us to input any value of b rather than use the value given outside the program.

```
func2[x_, y_] := (b = y + 1; x^2 + b)
```

One potential problem that can occur when user-defined functions are written this way is that values assigned to symbols outside the program are changed to the values used for these symbols in the program once it is run. For example, after entering

```
func2[a, 5]
```

$6 + a^2$

the value of b that was previously assigned in the kernel to be 23 is now changed.

```
b
```

6

The way to prevent this change in the values assigned to a symbol in the kernel is to use local variables. You can do this with Module, which allows you to declare variables as local. When you do this, *Mathematica* knows that when you refer to an x in a Module that has a local variable x, you're talking about the local version of x, and not any other version that might exist in the kernel (that also means that you won't be able to refer to your local value of x outside the Module).

Of course, you need to tell Module which variables you wish to localize. The first argument to Module should be a list of such variables. The second argument is the expression (typically a CompoundExpression) to evaluate. For example we can rewrite *func2* using Module

```
func3[x_, y_] := Module[{b}, b = y + 1; x^2 + b]
```

All of the programs in this book are written using the Module construction.

A.3.3 Ordering Functions

Just as there can be more than one transformation rule governing how different types of expressions should be altered, there can also be more than one function definition that uses the same head. Which definition applies to a particular set of arguments depends on which pattern the arguments match. This is the crux of most of the function definitions in this book.

For example, here is the definition of a function that returns True if its argument is 0, and False otherwise.

```
zeroQ[0] := True
zeroQ[_] := False
```

We see that this does what we want.

```
{zeroQ[20], zeroQ[0], zeroQ[4 - 5 + 1]}
```

{False, True, True}

It is important to note that when you give multiple function definitions for the same symbol, *Mathematica* attempts to put them in order from the most specific to the most general. When you then use the function, *Mathematica* tries the definitions in that order until it finds one whose left-hand side matches the given arguments. If the arguments don't match any of the left-hand sides, the expression is returned unevaluated.

In the example, *Mathematica* first checks to see if the most specific rule applies — that is, if the argument to *zeroQ* is a 0. If it is, True is returned. Otherwise, if the argument is not zero, it goes on to check if the second rule applies. The pattern for the second definition is very general, matching any expression and returning False in that case. So since the second definition only applies in cases where the argument is not zero, we don't need an explicit Condition requiring that it be nonzero. To make the second definition even more general, use BlankNullSequence to match not just any expression, but any sequence of zero or more expressions.

Thus, we can simplify function definitions by understanding the order in which *Mathematica* uses them.

Mathematica will show you how it has ordered multiple function definitions if you use the syntax ? followed by the function name.

```
?zeroQ
```

Global`zeroQ

```
zeroQ[0] := True

zeroQ[_] := False
```

It is always a good idea to periodically check *?yourFunction* to see if *Mathematica* has the definitions in the order in which you think they should be. More specific function definitions will automatically be put higher up on the list, but in those cases where *Mathematica* cannot tell the difference between how specific two function definitions are, it will put them in the order you have given.

More complicated examples of this can be found on nearly every page of this book, but one rule set that appears over and over is the walk rules (which are explained in Chapter 1). Here are the first few walk rules.

```
RND := Random[Integer, {1, 4}];
walk[{1,a___},0,_,_,_,{4,___},_,_,_,_,_,_,_] := {RND,a}
walk[{1,a___},0,_,_,_,_,_,_,{2,___},_,_,_,_] := {RND,a}
walk[{1,a___},0,_,_,_,_,_,_,_,{3,___},_,_,_] := {RND,a}
walk[{1,a___},0,_,_,_,_,_,_,_,_,_,_,_] := 0
```

The four preceding definitions specify how to transform expressions with head walk that have one as the first element of the first argument, and a zero as the second argument. Notice that the last definition is more general than the first three, therefore it won't apply unless all of the first three fail to apply.

```
{walk[{1, 3}, 0, 0, 0, 0, 0, 0, 0, {2}, 0, 0, 0, 0],
 walk[{1, 3}, 0, 0, 0, 0, 0, 0, 0, 0, 0, 0, 0, 0],
 walk[0, 0, 0, 0, 0, 0, 0, 0, 0, 0, 0, 0, 0]}

{{3, 3}, 0, walk[0, 0, 0, 0, 0, 0, 0, 0, 0, 0, 0, 0, 0]}
```

If there is no definition whose pattern matches the given arguments, then *Mathematica* returns the expression unevaluated, as in the preceding case of walk applied to 13 zeros. In the simulation programs in this book, a sequence of arguments that causes a function to return unevaluated is an oversight that needs to be fixed. This is something to keep an eye out for when writing your own simulations.

It is important to be careful in the labeling of patterns in function definitions. When the left-hand sides of two definitions are identical, having the same head and pattern labels, only the most recently entered definition will be kept in the kernel. But if the heads are the same and the pattern labels are different, then both defintions will be in the kernel which can result in the wrong definition being used.

A.3.4 Clearing Functions

When you make function definitions or assignments to variables, those definitions remain in the kernel until you quit the current *Mathematica* session. This is not always desirable: sometimes you want to remove a definition you've already made so that you can define it to mean something else.

There are many ways to deal with this; we mention one. You can use Clear[symbol] to remove any definitions or rules you have attached to the symbol. The following input removes all the function definitions associated with zeroQ.

```
?zeroQ

Global`zeroQ

zeroQ[0] := True

zeroQ[_] := False

Clear[zeroQ]; ?zeroQ

Global`zeroQ
```

Clearing variable names and function names is a good habit to get into when debugging your simulation programs. It is sometimes easy to overlook a definition that will cause your program to fail, even though you seem to be working on an entirely different function.

A.3.5 Anonymous Functions

We've discussed *Mathematica*'s ability to let users name their own functions. However, it is not always necessary to give user-defined functions a name. Functions without names are called *anonymous functions*. The usefulness of anonymous functions becomes apparent shortly.

How does one construct a function with no name? Recall the definition of *func1*:

```
?func1

Global`func1

func1[x_] := x^2 + 1
```

To write *func1* without a name we need only rewrite the function using the built-in Function command. The first argument to Function is a list of the pattern labels used in the function definition and the second argument is the left-hand side of the function definition.

```
Function[{x}, x^2 + 1]
```

$Function[\{x\}, x^2 + 1]$

We can then use this function just as we would use *func1*, applying it in exactly the same way.

```
Function[{x}, x^2 + 1][expr]
```

$1 + expr^2$

An even more common way to write anonymous functions leaves the arguments anonymous as well as the function name. In this case, the variables are replaced by # or #1 for the first argument, #2 for the second argument, and so on. Since there are no named variables, we no longer need the first argument to Function. The Function wrapper itself can be replaced by (...)&. So this same function can be rewritten again as

```
(#^2 + 1)&
```

Once more, the function is applied just as func was, with square brackets

```
(#^2 + 1)&[expr]
```

$1 + expr^2$

Care should be taken when applying anonymous functions to other anonymous functions, which is done throughout this book to combine successive function applications into a single command. When interpreting such functions, always read from the inside out, associating the # symbols with the nearest enclosing &. For instance, the following nested anonymous function first squares *expr*, and then adds one to the result.

```
(# + 1)&[(#^2)&[expr]]
```

$1 + expr^2$

When should you use an anonymous function and when should you use a named function? It depends heavily on what you wish to do. The difference between using a function definition and its anonymous function equivalent is that the latter does not go into the kernel. Hence, if you need to use a function in more than one place, you should probably name it so that the definition of the function need not be written more than once. On the other hand, anonymous functions can be convenient to use with functions such as PatternTest, that take a function as an argument.

```
532 /. x_?(# > 500&) → x-1
```

531

Notice that the anonymous predicate function is written slightly differently from other anonymous functions, with the right parenthesis moved to the right of the ampersand (...&).

A.4 Evaluation

The following discussion is somewhat technical and can be skipped upon first reading this appendix. However, in order to fully master the use of *Mathematica*, it is important to understand how *Mathematica* processes the inputs it receives.

A.4.1 Evaluation of Expressions

Mathematica is a term-rewriting system (TRS). Whenever an expression is entered, it is evaluated by term rewriting using rewrite rules. These rules consist of two parts: a pattern on the left-hand side (lhs) and a replacement text on the right-hand side (rhs). When the lhs of a rewrite rule is found to pattern-match part of the expression, that part is replaced by the rhs of the rule, after substituting values in the expression that match labeled blanks in the pattern into the rhs of the rule. Evaluation then proceeds by searching for further matching rules until no more are found.

The evaluation process in *Mathematica* can be easily understood with the following analogy: think of your experiences with using a handbook of mathematical formulas. In order to solve an integral, you consult the handbook which contains formulas consisting of a left-hand side (lhs) and a right-hand side (rhs), separated by an "equals" sign. You look for an integration formula in the handbook whose left-hand side has the same form as your integral.

Note: Athough no two formulas in the handbook have the identical lhs, there may be several whose lhs have the same form as your integral (eg., one lhs might have specific values in the integration limits or in the integrand, and another lhs has unspecified (dummy) variables for these quantities). When this happens, you use the formula whose lhs gives the closest fit to your integral.

You then replace your integral with the right-hand side of the matching lhs and you substitute the specific values in your integral for the corresponding variable symbols in the rhs. Finally, you look through the handbook for formulas (e.g., trigonometric identities or algebraic manipulation) that can be used to change the answer further.

This depiction provides an excellent description of the *Mathematica* evaluation process. However, the application of the term-rewriting process to a *Mathematica* expression requires a bit more discussion because a *Mathematica* expression consists of parts, a head and zero or more arguments which are themselves expressions.

$$expr0[\ expr1,\ expr2,\ .....,\ exprn\ ]$$

It is therefore necessary to understand the order in which the various parts of an expression are evaluated by term rewriting.

The implementation of the evaluation procedure is (with a few exceptions) straightforward.

- If the expression is a number or a string, it isn't changed.

- If the expression is a symbol, it is rewritten if there is an applicable rewrite rule in the global rule base; otherwise, it is unchanged.

- If the expression is not a number, string, or symbol, its parts are evaluated in a specific order.

- The head of the expression is evaluated.

- The arguments of the expression are evaluated from left to right in order. An exception to this occurs when the head is a symbol with a Hold attribute (e.g., HoldFirst, HoldRest, or HoldAll), so that some of its arguments are left in their unevaluated forms (unless they, in turn, have the head Evaluate), for example, the Set or SetDelayed function.

- After the head and arguments of an expression are each completely evaluated, the expression consisting of the evaluated head and arguments is rewritten, after making any necessary changes to the arguments based on the Attributes (such as Flat, Listable, Orderless) of the head, if there is an applicable rewrite rule in the global rule base.

- After carrying out the previous steps, the resulting expression is evaluated in the same way and then the result of that evaluation is evaluated, and so on until there are no more applicable rewrite rules.

The details of the term-rewriting process are:

- part of an expression is pattern-matched by the lhs of a rewrite rule;

- the values that match labeled blanks in the pattern are substituted into the rhs of the rewrite rule and evaluated; and

- the pattern-matched part of the expression is replaced with the evaluated result.

Appendix B

List and Matrix Essentials

In this appendix, we discuss some *Mathematica* functions that are useful for manipulating and analyzing lists and their contents. The techniques used here focus on examples from this book.

B.1 List Manipulation

The *Mathematica* function List is used to gather expressions together, similar to the way an egg carton can be used to gather eggs together, but can also be used to hold paper clips or rubber bands. We've already seen simple examples of Lists, which are just sequences of expressions wrapped in "curly brackets" ("{" and "}"). The FullForm of one of those examples shows that it is an expression with the head List.

```
FullForm[{Head[Plus], Head["Simulating Society"], Head[51]}]
```

```
List[Symbol, String, Integer]
```

Any expression can be an argument to List, including another List. Thus, List can be used to store linear one-dimensional lists, two-dimensional matrices of expressions, matrices of arbitrary dimensions, or a mixture of lists and other expressions. This allows flexible grouping and storing of data, as well as myriad ways of rearranging and analyzing those data. Manipulating matrices is discussed in the next section, but here we focus on one-dimensional lists. Throughout this section, we use the following example.

```
person = {3, 1, 1, {2, 3, 4}, 4}
```

{3, 1, 1, {2, 3, 4}, 4}

This list is taken from Chapter 1, where it represents a typical individual in our simulations. Each element of the list can be interpreted as storing a piece of information about the individual. We interpret the preceding individual in the following way.

- The first element in the list (3) represents the direction in which the individual is facing, which is used to determine his movement at a given moment.

- The second element (1) represents the name of the individual, allowing us to distinguish him from other individuals during the simulation, or keep track of him over time.

- The third element (1) determines how the individual interacts with others at a given moment. For instance, a value of zero may mean that the individual's behavior is bad, whereas a value of one may mean that his behavior is good.

- The fourth element of the list is itself a list, representing either fixed or changeable values or attitudes of the individual. This list is referred to in the book as the individual's *meme* list. Each position in the meme list holds the numerical value of a particular attribute. For instance, the first element of the list might refer to race, with different numbers corresponding to different races. Other possible interpretations include religion, IQ, height, or political loyalty.

- The last element in the list (4) represents the accumulated resources of the individual. These resources can be material in nature (such as an accumulation of money or property) or nonmaterial (such as psychological or social distinctions).

In the matrix manipulation section later, we discuss how to generate a random individual with this given format, and also how to generate a society with a given population density of such individuals.

B.1.1 Table

One of the most basic list creation commands available in *Mathematica* is the Table command. There are many ways to call the Table command, but the most basic one creates a list of n identical elements. We use $n = 5$ for a concrete example.

```
n = 5; Table[0, {n}]
```

{0, 0, 0, 0, 0}

The $\{n\}$ is called an *iterator* for the Table. Table can generate a list of lists — that is, a list whose elements are themselves lists — by specifying more than one iterator. Here is an example of a list that has n elements, each of which is a list of n zeros. (This is an "n by n" square matrix.)

```
n = 5; Table[0, {n}, {n}]
```

{{0, 0, 0, 0, 0}, {0, 0, 0, 0, 0}, {0, 0, 0, 0, 0},
{0, 0, 0, 0, 0}, {0, 0, 0, 0, 0}}

The primary use of Table in our simulations is to create a society (a two-dimensional matrix containing individuals and empty sites) with certain characteristics. We discuss this more fully in the following matrix manipulation section.

B.1.2 Extracting Information from a List

There are a number of functions that allow us to examine a list. For example, we can find out the number of elements in a list using the built-in Length function. Recall our previously defined person.

```
person
```

{3, 1, 1, {2, 3, 4}, 4}

This list has five elements, and so Length[person] returns 5.

```
Length[person]
```

5

There are also ways to extract a particular element or elements of a list. The functions First and Last extract the first or last element of a list, respectively, and Rest returns a list with the first element dropped. For instance, we can extract the resource level of our individual and the direction that the individual is facing with the following bit of code.

```
{Last[person], First[person]}
```

{4, 3}

To extract the individual's meme list, which is neither the first nor last element of the list, we need to use the built-in Part function. To extract the ith element of a list *lis* (or of any expression), we ask for Part[*lis*, i]. This can also be written *lis*[[i]]. In particular, the meme list is the fourth element of the individual's list.

```
person[[4]]
```

{2, 3, 4}

If you give a negative number as the second argument to Part, the positions are counted starting at the end of the list. In the case of our individual, the fourth element in the list is the same as the minus-second element.

```
person[[-2]]
```

```
{2, 3, 4}
```

We can take this same notation one step further to extract the second element of the meme list. Extracting expressions from arbitrary matrices is discussed in the matrix manipulation section.

In cases where you know which element of the list you want to extract, Part works well. However, there are times when you need to figure out which elements satisfy some condition that you specify, or match a pattern that you give. For example, the Position function takes an expression and a pattern, and returns all positions within the list where the pattern finds a matching expression. For example, we can search our sample individual for all occurrences of the number 1.

```
person
```

```
{3, 1, 1, {2, 3, 4}, 4}
```

```
Position[person, 1]
```

```
{{2}, {3}}
```

So the number 1 appears as the second element of person, and again as the third element. We show why there is an extra set of list braces around each position in the matrix manipulation section. We also show how to interpret the result of Position when it locates an expression that is nested within an element of the list, as in this example.

```
Position[person, 3]
```

```
{{1}, {4, 2}}
```

Here, we use Position to begin understanding the contents of the individual's meme list, obtaining a list of positions that hold values less than four.

```
Position[person[[4]], _?(# < 4&)]
```

```
{{1}, {2}}
```

B.1.3 Select, Cases, and Count

Three other functions for extracting information out of lists are Select, Cases, and Count. Select uses the list as its first argument, and a (predicate) function as the second argument (usually an anonymous function). The output is a list that contains only those elements that cause the function to return True. Note that this is different from Position, which only returned a list of positions and not a list of elements. For instance, recall the example individual.

```
person
```

$\{3, 1, 1, \{2, 3, 4\}, 4\}$

Even without knowing where the meme list is in the individual's list, we can use Select to obtain a list of all the elements that have head List.

```
Select[person, (Head[#] === List &)]
```

$\{\{2, 3, 4\}\}$

If we are looking only for individuals whose meme list satisfies some condition, such as those whose first element is a 5, it's possible that Select will find no elements. In this case, it returns an empty list.

```
Select[person, (MatchQ[#, {5, ___}]&)]
```

$\{\}$

The function Cases is also used to extract elements from a list, but it does so using pattern-matching. Any element of the list that matches the specified pattern is returned.

```
Cases[person, _List]
```

$\{\{2, 3, 4\}\}$

The second argument to Cases can also be a replacement rule that will transform expressions that are picked out of the list in whatever way you specify. For instance, you could extract every sublist from a given list, computing each sublist's average as you do so. The calculation that computes the average uses the Apply function, which is discussed later in this appendix.

```
Cases[person, x_List :> Apply[Plus, x]/Length[x]]
```

$\{3\}$

The function Count works much like Cases in that it uses pattern-matching to obtain a result. However, instead of returning a list of elements, it simply returns the number of elements found to match the given pattern.

```
Count[person, _List]
```

1

B.1.4 Rearranging Lists

Rearranging lists is something that can be done easily and elegantly in *Mathematica*. For instance, a common way to reorganize a list is to sort it. The built-in Sort function does this.

```
lis = {1, -2, 3, -4, 5, -6, 0}; Sort[lis]
```

{-6, -4, -2, 0, 1, 3, 5}

By default, Sort orders elements using the built-in Less (<) operator. However, we can provide any function of two arguments to be used as a sorting function by specifying it as the second argument to Sort. That sorting function is applied to pairs of elements from the list to rearrange them into the new order. For instance, this sorting function puts numbers with a larger absolute value first.

```
Sort[lis, (Abs[#1] > Abs[#2] &)]
```

{-6, 5, -4, 3, -2, 1, 0}

In the next section we show how Sort can be used to obtain from a simulation a list of individuals who have cultural values or other attributes in common.

Also useful is the ability to rotate a list, effectively moving either the first or the last element to the other end of the list. The functions that do this are RotateLeft and RotateRight, respectively. These functions also take a second argument that dictates how many elements should be rotated in this way. A negative second argument is interpreted as a rotation in the opposite direction.

```
person
```

{3, 1, 1, {2, 3, 4}, 4}

```
{RotateLeft[person], RotateRight[person, 3]}
```

{{1, 1, {2, 3, 4}, 4, 3}, {1, {2, 3, 4}, 4, 3, 1}}

B.2 Matrix Manipulation

We've used the basic list manipulation functions in conjunction with one-dimensional lists. We now use them for more complicated nested lists. There are additional features of most of these functions to allow one to handle such nested lists easily and efficiently.

For the remainder of this section, we use the following nested list to demonstrate the majority of the functions.

```
society = {{0, 0, 0, {3, 1, 1, {2, 3, 4}, 4}}, {0, {3, 2, 1,
{4, 3, 3}, 3}, {2, 3, 1, {4, 3, 4}, 1}, {2, 4, 1, {5, 3, 1},
3}}, {0, {1, 5, 1, {5, 3, 5}, 2}, 0, {1, 6, 0, {2, 1, 5},
4}}, {0, 0, 0, {4, 7, 1, {4, 5, 5}, 3}}};
```

Mathematica can display this formatted as a two-dimensional matrix (instead of as a list of lists) if we ask for the TraditionalForm of the matrix.

```
TraditionalForm[society]
```

$$\begin{pmatrix} 0 & 0 & 0 & \{3, 1, 1, \{2, 3, 4\}, 4\} \\ 0 & \{3, 2, 1, \{4, 3, 3\}, 3\} & \{2, 3, 1, \{4, 3, 4\}, 1\} & \{2, 4, 1, \{5, 3, 1\}, 3\} \\ 0 & \{1, 5, 1, \{5, 3, 5\}, 2\} & 0 & \{1, 6, 0, \{2, 1, 5\}, 4\} \\ 0 & 0 & 0 & \{4, 7, 1, \{4, 5, 5\}, 3\} \end{pmatrix}$$

This *society* matrix is taken from Chapter 1, and is exactly the kind of matrix or *lattice* that appears throughout this book. In particular, it is a matrix that has the same number of rows (4) as columns (4), where the elements of the rows or *sites* are either empty or occupied by an individual. Empty sites are represented by a zero, and occupied sites contain a list in which is stored all the information we have about the individual. Analysis of a particular individual's list was the basis for the previous section. Notice that the example individual from the previous section is in the first row of this society example.

B.2.1 Building Society

The example matrix defined previously was actually created by *Mathematica* using a set of instructions in Chapter 1. In this section, we describe those instructions and show exactly how that matrix was created.

Random and SeedRandom

Mathematica has a built-in function Random that can generate various sorts of random numbers. Random[] returns a random real number between 0 and 1. Random[*type*, {*min*, *max*}] generates a random number of the specified *type* between *min* and *max*. For example, here's an important function that is used throughout the book.

```
RND := Random[Integer, {1, 4}]
Table[RND, {10}]
```

```
{3, 2, 1, 4, 2, 2, 2, 1, 2, 4}
```

This RND function returns a random integer between one and four inclusive, which is used in deciding the direction an individual faces and moves.

As with any computer language, *Mathematica* does not generate truly random numbers; rather, it uses some algorithm to generate a sequence of numbers that seem random. (For implementation details regarding *Mathematica*'s random number generator, see the *Mathematica* book.) It is sometimes useful to be able to generate the same sequence of random numbers over and over. For instance, this would make it possible to begin a simulation with the same random initial lattice configuration each time it was run.

One way to do this is by using SeedRandom, which supplies the pseudorandom number generation algorithm with a seed or starting value. At any point in any kernel session, you can then put the random number generator into the same state by providing the same value to SeedRandom. For example, notice that the outputs for the next two inputs are identical.

```
SeedRandom[0]; Table[RND, {10}]
```

{4, 2, 4, 1, 3, 3, 1, 2, 3, 4}

```
SeedRandom[0]; Table[RND, {10}]
```

{4, 2, 4, 1, 3, 3, 1, 2, 3, 4}

Floor

One of the requirements for every simulation in this book is to be able to specify the initial population density as a real number between zero and one. This can be done through the use of Random and Floor. Floor is a numerical function that returns the largest integer less than its argument. So, for example, Floor of a random real number between zero and one will always return zero.

```
Floor[Random[]]
```

0

Important for creating populations with a given density is the fact that Floor of a random real number between p and $1 + p$ will yield 1 about p percent of the time, and 0 in the remaining cases. We can generate a random real between p and $1 + p$ either by using Random[Real, $\{p, 1 + p\}$], or by adding p to a random real number between 0 and 1, as in p + Random[]. The latter is more efficient and easier to read.

Thus, we can create a list that contains approximately 80% ones and 20% zeros with the following code.

```
p = 0.8;
SeedRandom[1]
Table[Floor[p + Random[]], {15}]
```

{1, 1, 1, 0, 1, 1, 1, 1, 0, 1, 1, 1, 1, 0, 1}

PreIncrement

Many of the models in this book require us to name the individuals in the system. The simplest way to do this is to name them consecutively, starting at 1 and ending at the number of individuals there are in the population. In general, though, we do not know the number of individuals beforehand; we only know the target population density. In order to keep track of the number of individuals, we use a *counter* variable along with the PreIncrement function. PreIncrement[*counter*], or ++*counter*, sets the new value of *counter* to be the current value of *counter* plus 1.

For instance, since each value of 1 in the previous example corresponds to a site occupied by a person, we can replace it with a name. This can be done with the replacement rule 1 :→ ++*counter*.

```
p = 0.8;
k = 0;
SeedRandom[1]
Table[Floor[p + Random[]], {15}] /. 1 :→ ++k

{1, 2, 3, 0, 4, 5, 6, 7, 0, 8, 9, 10, 11, 0, 12}
```

Building Society

Recall that the Table function can take more than one iterator, thereby specifying a table in more than one dimension. With this information, and the techniques just developed, we can generate a lattice that has the following properties.

- The lattice has four rows and four columns ($n = 4$).

- The population density is 25% ($p = 0.25$). Empty sites are represented by zeros.

We can see just these two properties without further restriction in the following lattice.

```
n = 4; p = 0.25;
SeedRandom[3]
Table[Floor[p + Random[]], {n}, {n}]

{{0, 0, 0, 1}, {0, 1, 1, 1}, {0, 1, 0, 1}, {0, 0, 0, 1}}
```

Now, suppose each individual on the lattice is represented by a list with five elements:

- a randomly chosen integer between 1 and 4, representing the direction the individual is facing;

- an integer between 1 and the total number of individuals on the lattice, representing the name of the individual;

- a zero or one denoting a badly or well-behaved individual, such that approximately 75% of the individuals are well-behaved ($g = 0.75$);

- a list of three integers ($s = 3$) randomly chosen between 1 and 5 ($m = 5$), representing the individual's meme list; and

- a randomly chosen integer between 1 and 4 ($r = 4$) representing the individual's starting resources.

We can create a system of such individuals by replacing every 1 with a list by using a suitably constructed replacement rule. Doing so completes the construction of the example *society* matrix given previously.

```
n = 4; p = 0.25; g = 0.75; m = 5; s = 3; r = 4;
k = 0;
SeedRandom[3]
RND := Random[Integer, {1, 4}]
newSociety = Table[Floor[p + Random[]], {n}, {n}] /.
  1 :> {RND, ++k, Floor[g + Random[]],
        Table[Random[Integer, {1, m}], {s}],
        Random[Integer, {1, r}]};
TraditionalForm[newSociety]
```

$$
\begin{pmatrix}
0 & 0 & 0 & \{3, 1, 1, \{2, 3, 4\}, 4\} \\
0 & \{3, 2, 1, \{4, 3, 3\}, 3\} & \{2, 3, 1, \{4, 3, 4\}, 1\} & \{2, 4, 1, \{5, 3, 1\}, 3\} \\
0 & \{1, 5, 1, \{5, 3, 5\}, 2\} & 0 & \{1, 6, 0, \{2, 1, 5\}, 4\} \\
0 & 0 & 0 & \{4, 7, 1, \{4, 5, 5\}, 3\}
\end{pmatrix}
$$

B.2.2 Extracting Information from Society

All the functions that work for the list *person* also work for the matrix *society*, although you need to be careful to make sure they're doing what you want. For example, the functions Length, First, Last, and Rest now will operate on the rows of the matrix.

```
First[society]
```

```
{0, 0, 0, {3, 1, 1, {2, 3, 4}, 4}}
```

The function Part that we used in the form *lis*[[*i*]] is general enough to extract elements out of any list. When dealing with a matrix, we can use Part to extract the elements of the list elements, as long as we specify the *index* of the elements we want. For example, if you wanted to extract the meme list of the individual named "1" in *society*, you would want the fourth element of the fourth element of the first element of the matrix.

```
society[[1]][[4]][[4]]
```

```
{2, 3, 4}
```

A more pleasing syntax for this combines the indices into a sequence and gets rid of extra brackets.

```
society[[1, 4, 4]]
```

{2, 3, 4}

Finally, on those occasions where you need to extract the head of an expression, you can refer to its zeroth part. This can be useful to make sure that you've extracted the correct expression.

```
society[[1, 4, 4, 0]]
```

List

Position, like Part, applies to matrices as well as lists. In fact, Position doesn't just analyze the elements (or rows) of the matrix, but it analyzes any expression that exists anywhere within the matrix. For example, we can extract the positions of all the empty sites in the matrix. However, asking for the position of all zeros in our example is not specific enough.

```
Position[society, 0]
```

{{1, 1}, {1, 2}, {1, 3}, {2, 1}, {3, 1}, {3, 3}, {3, 4, 3},
  {4, 1}, {4, 2}, {4, 3}}

Notice that this returns position {3, 4, 3} which does not correspond to an empty site, but rather is in the midst of an individual's list. The zero, being the third element of that individual's list, indicates that he is behaving badly.

```
society[[3, 4]]
```

{1, 6, 0, {2, 1, 5}, 4}

We can restrict Position to returning only those positions of the matrix that are specified by two numbers by providing a third argument or *level specification* of {2}.

```
Position[society, 0, {2}]
```

{{1, 1}, {1, 2}, {1, 3}, {2, 1}, {3, 1}, {3, 3}, {4, 1},
  {4, 2}, {4, 3}}

We discuss level specifications in somewhat more detail later on in this appendix.

Finally notice that Position confirms that the list *person* is actually a member of the matrix *society*.

```
Position[society, person]
```

{{1, 4}}

B.2.3 Cases and Count

Armed with a level specification, Cases and Count can be used to extract information from the matrix quite easily. For example, Cases can give us a list of the individuals that are present in *society*.

```
Cases[society, _List, {2}]
```

```
{{3, 1, 1, {2, 3, 4}, 4}, {3, 2, 1, {4, 3, 3}, 3},
 {2, 3, 1, {4, 3, 4}, 1}, {2, 4, 1, {5, 3, 1}, 3},
 {1, 5, 1, {5, 3, 5}, 2}, {1, 6, 0, {2, 1, 5}, 4},
 {4, 7, 1, {4, 5, 5}, 3}}
```

Recall that Cases can take a transformation rule as its second argument. Here is an example of a transformation rule that extracts the meme list of all the good individuals in *society*.

```
Cases[society, {_, _, 1, x_, _} :→ x, {2}]
```

```
{{2, 3, 4}, {4, 3, 3}, {4, 3, 4}, {5, 3, 1}, {5, 3, 5}, {4, 5, 5}}
```

Similarly, Count can tell us how many individuals have resources that are less than some given level, provided we supply the level specification.

```
Count[society, {_, _, _, _, r_} /; r < 3, {2}]
```

```
2
```

When a simulation run is finished, we generally have a list of matrices, one for each time step during which the matrix was updated, instead of a single matrix. In those cases, a deeper level specification can be used to extract information. The first level of such a list would consist of the matrices themselves, the second level would be the rows, and the third level would correspond to the site values; thus, a level specification of {3} can be used to analyze just the site values. In this way, we can extract information (such as the position of an individual or the population's average resource level) from each matrix and analyze it as a function of time.

B.2.4 Rearranging Matrices

The optional predicate function given as a second argument to Sort can be used to sort lists of lists by any of their elements. By default, Sort arranges the rows of a matrix based on the first element of the row. If the first elements are identical, it moves to the second element, and so on. For example, here is a list of the individuals in *society* sorted by the direction they are facing.

```
Sort[Cases[society, _List, {2}]]
```

```
{{1, 5, 1, {5, 3, 5}, 2}, {1, 6, 0, {2, 1, 5}, 4},
 {2, 3, 1, {4, 3, 4}, 1}, {2, 4, 1, {5, 3, 1}, 3},
 {3, 1, 1, {2, 3, 4}, 4}, {3, 2, 1, {4, 3, 3}, 3},
 {4, 7, 1, {4, 5, 5}, 3}}
```

In other applications, it may be necessary to sort these same individuals by their resource levels. We can do this by instructing Sort to base the order on the last element of the sublists.

```
Sort[Cases[society, _List, {2}], (Last[#1] < Last[#2]&)]
```

```
{{2, 3, 1, {4, 3, 4}, 1}, {1, 5, 1, {5, 3, 5}, 2},
 {4, 7, 1, {4, 5, 5}, 3}, {2, 4, 1, {5, 3, 1}, 3},
 {3, 2, 1, {4, 3, 3}, 3}, {1, 6, 0, {2, 1, 5}, 4},
 {3, 1, 1, {2, 3, 4}, 4}}
```

Instead of a level specification, RotateLeft and RotateRight can take a list as their second argument that tells what to do to *successive levels* of the list. (The first level of a list is all those elements that can be accessed with one number via Part, as in *lis*[[*i*]]. The second level of a list is those elements that can be accessed with two numbers via Part, as in *lis*[[*i, j*]], etc.) For example, the following RotateLeft moves the first column to the last column in the matrix.

```
TraditionalForm[society]
```

$$
\begin{pmatrix}
0 & 0 & 0 & \{3, 1, 1, \{2, 3, 4\}, 4\} \\
0 & \{3, 2, 1, \{4, 3, 3\}, 3\} & \{2, 3, 1, \{4, 3, 4\}, 1\} & \{2, 4, 1, \{5, 3, 1\}, 3\} \\
0 & \{1, 5, 1, \{5, 3, 5\}, 2\} & 0 & \{1, 6, 0, \{2, 1, 5\}, 4\} \\
0 & 0 & 0 & \{4, 7, 1, \{4, 5, 5\}, 3\}
\end{pmatrix}
$$

```
TraditionalForm[RotateLeft[society, {0, 1}]]
```

$$
\begin{pmatrix}
0 & 0 & \{3, 1, 1, \{2, 3, 4\}, 4\} & 0 \\
\{3, 2, 1, \{4, 3, 3\}, 3\} & \{2, 3, 1, \{4, 3, 4\}, 1\} & \{2, 4, 1, \{5, 3, 1\}, 3\} & 0 \\
\{1, 5, 1, \{5, 3, 5\}, 2\} & 0 & \{1, 6, 0, \{2, 1, 5\}, 4\} & 0 \\
0 & 0 & \{4, 7, 1, \{4, 5, 5\}, 3\} & 0
\end{pmatrix}
$$

Rotations of this kind are indispensable for creating the neighborhood of a site, which is discussed in detail in the next section.

Another method of rearranging a matrix is to exchange the rows with the columns. Transpose is the function that does this. Notice that the first row of *society* becomes the first column of Transpose[*society*].

```
TraditionalForm[Transpose[society]]
```

$$
\begin{pmatrix}
0 & 0 & 0 & 0 \\
0 & \{3, 2, 1, \{4, 3, 3\}, 3\} & \{1, 5, 1, \{5, 3, 5\}, 2\} & 0 \\
0 & \{2, 3, 1, \{4, 3, 4\}, 1\} & 0 & 0 \\
\{3, 1, 1, \{2, 3, 4\}, 4\} & \{2, 4, 1, \{5, 3, 1\}, 3\} & \{1, 6, 0, \{2, 1, 5\}, 4\} & \{4, 7, 1, \{4, 5, 5\}, 3\}
\end{pmatrix}
$$

This is important in many different applications. For example, here are two ways to extract the resource level from every individual in *society*. The first method uses Cases, and the second uses Transpose.

```
Cases[society, x_List :> Last[x], {2}]
```

{4, 3, 1, 3, 2, 4, 3}

```
Last[Transpose[Cases[society, _List, {2}]]]
```

{4, 3, 1, 3, 2, 4, 3}

B.3 Higher Order Functions

B.3.1 Map, MapThread, and Neighborhoods

A function can be applied to each element in a list using Map. This is useful for transforming all the elements in a list at once, instead of looping over the list. For example, this applies the undefined function h to each of the elements of the given list.

```
Map[h, {2, 1, 3}]
```

{h[2], h[1], h[3]}

Map[*function*, *expr*], can also be written *function* /@ *expr*.

Like Cases and Count, Map can take a level specification. For instance, if you wanted to transform society by some function h applied to each individual or empty site, you could do that using Map with a level specification of {2}. We can compare the results of mapping the function h over *society*, and mapping it over *society* at level {2}.

```
TraditionalForm[society]
```

$$\begin{pmatrix} 0 & 0 & 0 & \{3, 1, 1, \{2, 3, 4\}, 4\} \\ 0 & \{3, 2, 1, \{4, 3, 3\}, 3\} & \{2, 3, 1, \{4, 3, 4\}, 1\} & \{2, 4, 1, \{5, 3, 1\}, 3\} \\ 0 & \{1, 5, 1, \{5, 3, 5\}, 2\} & 0 & \{1, 6, 0, \{2, 1, 5\}, 4\} \\ 0 & 0 & 0 & \{4, 7, 1, \{4, 5, 5\}, 3\} \end{pmatrix}$$

Notice that TraditionalForm uses parentheses instead of square brackets for function application, to conform to standard mathematical notation.

```
TraditionalForm[Map[h, society]]
```

{h({0, 0, 0, {3, 1, 1, {2, 3, 4}, 4}}),
  h({0, {3, 2, 1, {4, 3, 3}, 3}, {2, 3, 1, {4, 3, 4}, 1}, {2, 4, 1, {5, 3, 1}, 3}}),
  h({0, {1, 5, 1, {5, 3, 5}, 2}, 0, {1, 6, 0, {2, 1, 5}, 4}}), h({0, 0, 0, {4, 7, 1, {4, 5, 5}, 3}})}

```
TraditionalForm[Map[h, society, {2}]]
```

$$\begin{pmatrix} h(0) & h(0) & h(0) & h(\{3,\ 1,\ 1,\ \{2,\ 3,\ 4\},\ 4\}) \\ h(0) & h(\{3,\ 2,\ 1,\ \{4,\ 3,\ 3\},\ 3\}) & h(\{2,\ 3,\ 1,\ \{4,\ 3,\ 4\},\ 1\}) & h(\{2,\ 4,\ 1,\ \{5,\ 3,\ 1\},\ 3\}) \\ h(0) & h(\{1,\ 5,\ 1,\ \{5,\ 3,\ 5\},\ 2\}) & h(0) & h(\{1,\ 6,\ 0,\ \{2,\ 1,\ 5\},\ 4\}) \\ h(0) & h(0) & h(0) & h(\{4,\ 7,\ 1,\ \{4,\ 5,\ 5\},\ 3\}) \end{pmatrix}$$

Related to Map is MapThread, which can be used to *thread* or *zip* together lists that have the same structure, so that corresponding elements are paired up as arguments to a function. We can see the syntax and even a simple example in its usage message. Notice that like Map, MapThread takes a level specification.

```
?MapThread
```

*MapThread[f, {{a1, a2, ... }, {b1, b2, ... }, ... }]
    gives {f[a1, b1, ...], f[a2, b2, ...], ... }.
    MapThread[f, {expr1, expr2, ... }, n] applies f to
    the parts of the expri at level n.*

For instance, if we wanted to apply the function h to the elements of *society* paired with the elements of RotateRight[*society*, {0,1}], we can do so with MapThread. To make the output more readable, we look only at the second row of the resulting matrix.

```
society[[2]]
```

```
{0, {3, 2, 1, {4, 3, 3}, 3}, {2, 3, 1, {4, 3, 4}, 1},
 {2, 4, 1, {5, 3, 1}, 3}}
```

```
MapThread[h, {society, RotateRight[society, {0,1}]}][[2]]
```

```
h[{0, {3, 2, 1, {4, 3, 3}, 3}, {2, 3, 1, {4, 3, 4}, 1},
  {2, 4, 1, {5, 3, 1}, 3}}, {{2, 4, 1, {5, 3, 1}, 3}, 0,
  {3, 2, 1, {4, 3, 3}, 3}, {2, 3, 1, {4, 3, 4}, 1}}]
```

Using the level specification for MapThread, we can pair off the elements of the rows instead of the rows themselves.

```
MapThread[h, {society, RotateRight[society, {0,1}]}, 2][[2]]
```

```
{h[0, {2, 4, 1, {5, 3, 1}, 3}], h[{3, 2, 1, {4, 3, 3}, 3}, 0],
 h[{2, 3, 1, {4, 3, 4}, 1}, {3, 2, 1, {4, 3, 3}, 3}],
 h[{2, 4, 1, {5, 3, 1}, 3}, {2, 3, 1, {4, 3, 4}, 1}]}
```

Notice that each element of this matrix is h applied to the corresponding element of *society*, together with its neighbor to the left or west. We use *wraparound* boundary conditions, so that the western neighbor of the first element in a row corresponds to the last element of the row. We can use other second arguments of RotateRight to pair off sites with their northern ({1, 0}), eastern ({0, -1}), or southern ({-1, 0}) neighbors, in all cases with wraparound boundary conditions.

In fact, we can perform all four rotations at once, to create the *vonNeumann* neighborhood of a site, which consists of that site together with its northern, eastern, southern, and western neighbors. Instead of applying this to our *society* example, let's look at a matrix with simpler elements.

```
mat = {{1, north, 2}, {west, site, east}, {3, south, 4}};
TraditionalForm[mat]
```

$$\begin{pmatrix} 1 & north & 2 \\ west & site & east \\ 3 & south & 4 \end{pmatrix}$$

We generate the matrix of neighborhoods like so.

```
ngbds = MapThread[List, {mat,
    RotateRight[mat, {1, 0}], RotateRight[mat, {0, -1}],
    RotateRight[mat, {-1, 0}], RotateRight[mat, {0, 1}]}, 2];
TraditionalForm[ngbds]
```

$$\begin{pmatrix} \{1, 3, north, west, 2\} & \{north, south, 2, site, 1\} & \{2, 4, 1, east, north\} \\ \{west, 1, site, 3, east\} & \{site, north, east, south, west\} & \{east, 2, west, 4, site\} \\ \{3, west, south, 1, 4\} & \{south, site, 4, north, 3\} & \{4, east, 3, 2, south\} \end{pmatrix}$$

Notice in particular the center element of the resulting lattice.

```
ngbds[[2, 2]]
```

```
{site, north, east, south, west}
```

It contains exactly the sequence of sites we wanted. In fact, every site of the *ngbds* lattice contains that position's vonNeumann neighborhood. Furthermore, there's nothing special about threading List over the matrix of rotations. We could just as easily have used a different function *h*.

```
ngbds = MapThread[h, {mat,
    RotateRight[mat, {1, 0}], RotateRight[mat, {0, -1}],
    RotateRight[mat, {-1, 0}], RotateRight[mat, {0, 1}]}, 2];
TraditionalForm[ngbds]
```

$$\begin{pmatrix} h(1, 3, north, west, 2) & h(north, south, 2, site, 1) & h(2, 4, 1, east, north) \\ h(west, 1, site, 3, east) & h(site, north, east, south, west) & h(east, 2, west, 4, site) \\ h(3, west, south, 1, 4) & h(south, site, 4, north, 3) & h(4, east, 3, 2, south) \end{pmatrix}$$

Notice that there is a lot of repetition of the RotateRight command in the preceding input. We can simplify this by mapping RotateRight over the various rotations. RotateRight with a second argument of $\{0, 0\}$ performs no rotation.

```
ngbds = MapThread[h, Map[RotateRight[mat, #]&,
        {{0, 0}, {1, 0}, {0, -1}, {-1, 0}, {0, 1}}], 2];
TraditionalForm[ngbds]
```

$$\begin{pmatrix} h(1, 3, north, west, 2) & h(north, south, 2, site, 1) & h(2, 4, 1, east, north) \\ h(west, 1, site, 3, east) & h(site, north, east, south, west) & h(east, 2, west, 4, site) \\ h(3, west, south, 1, 4) & h(south, site, 4, north, 3) & h(4, east, 3, 2, south) \end{pmatrix}$$

We can gather the preceding commands into a new function definition, *vonNeumann*, that applies a given function to a given matrix of neighborhoods.

```
vonNeumann[func_, lat_] :=
  MapThread[func, Map[RotateRight[lat, #]&,
       {{0, 0}, {1, 0}, {0, -1}, {-1, 0}, {0, 1}}], 2];
```

Along these same lines, we can rotate the given matrix in other ways to get different neighborhoods. Two that occur throughout the book are the *Moore* neighborhood, and the *Gaylord–Nishidate* (or *GN*) neighborhood. The Moore neighborhood of a site consists of that site along with its north, east, south, west, northeast, southeast, southwest, and northwest neighbors. The GN neighborhood adds to these the sites to the north, south, east, and west that are two sites away from the site in question.

```
Moore[func_, lat_] :=
  MapThread[func, Map[RotateRight[lat, #]&,
       {{0, 0}, {1, 0}, {0, -1}, {-1, 0}, {0, 1},
        {1, -1}, {-1, -1}, {-1, 1}, {1, 1}}], 2];

GN[func_, lat_] :=
  MapThread[func, Map[RotateRight[lat, #]&,
       {{0, 0}, {1, 0}, {0, -1}, {-1, 0}, {0, 1},
        {1, -1}, {-1, -1}, {-1, 1}, {1, 1}, {2, 0},
        {0, -2}, {-2, 0}, {0, 2}}], 2];
```

B.3.2 Nest and NestList

When running the simulations in this book, all sites of the lattice are updated during the course of one time step. The resulting lattice is then updated in the same way, and then the result of that is updated, and so forth. This repeated function application can be done in *Mathematica* using the Nest or NestList functions. NestList[h, *expr*, n] applies h to *expr*, then applies h to the result, and so on, n times, and then returns a list of all the n + 1 resulting expressions, beginning with *expr*.

```
NestList[h, expr, 3]
```

```
{expr, h[expr], h[h[expr]], h[h[h[expr]]]}
```

Nest does the same calculations as NestList, but just returns the result of the repeated function application, and leaves out all the intermediate results. In effect, it returns the last element of NestList.

```
Nest[h, expr, 3]
```

```
h[h[h[expr]]]
```

There are also some memory savings that Nest exhibits over NestList. We discuss these in the next appendix.

For a more interesting example, recall that we can apply any function to the vonNeumann neighborhood of every site in the lattice simultaneously. Let's define one such function, *increaseResources*, which adds to the resource level of an individual a number equal to the number of good individuals other than themselves that populate their vonNeumann neighborhood. (Recall that good individuals are those that have a 1 as the third element of their list.) It should leave empty sites alone. This function might be called a *group effect*, as the results depend on each individual's neighborhood as a whole. We can define this function with the following two rules.

```
Clear[increaseResources]

increaseResources[{a_, b_, c_, d_, r_}, w_, x_, y_, z_] :=
    {a, b, c, d, r + Count[{w, x, y, z}, {_, _, 1, _, _}]};
increaseResources[0, _, _, _, _] := 0;
```

We can apply this function "by hand" in the traditional manner.

```
TraditionalForm[society]
```

$$\begin{pmatrix} 0 & 0 & 0 & \{3, 1, 1, \{2, 3, 4\}, 4\} \\ 0 & \{3, 2, 1, \{4, 3, 3\}, 3\} & \{2, 3, 1, \{4, 3, 4\}, 1\} & \{2, 4, 1, \{5, 3, 1\}, 3\} \\ 0 & \{1, 5, 1, \{5, 3, 5\}, 2\} & 0 & \{1, 6, 0, \{2, 1, 5\}, 4\} \\ 0 & 0 & 0 & \{4, 7, 1, \{4, 5, 5\}, 3\} \end{pmatrix}$$

```
TraditionalForm[vonNeumann[increaseResources, society]]
```

$$\begin{pmatrix} 0 & 0 & 0 & \{3, 1, 1, \{2, 3, 4\}, 6\} \\ 0 & \{3, 2, 1, \{4, 3, 3\}, 5\} & \{2, 3, 1, \{4, 3, 4\}, 3\} & \{2, 4, 1, \{5, 3, 1\}, 5\} \\ 0 & \{1, 5, 1, \{5, 3, 5\}, 3\} & 0 & \{1, 6, 0, \{2, 1, 5\}, 6\} \\ 0 & 0 & 0 & \{4, 7, 1, \{4, 5, 5\}, 4\} \end{pmatrix}$$

To apply this function 10 times, we can simply apply the function *vonNeumann*[*increaseResources*, #]& to the matrix, and then apply that same function to the result, and so on, 10 times. Nest returns the result of the final function application.

```
TraditionalForm[
    Nest[vonNeumann[increaseResources, #]&, society, 10]]
```

$$\begin{pmatrix} 0 & 0 & 0 & \{3, 1, 1, \{2, 3, 4\}, 24\} \\ 0 & \{3, 2, 1, \{4, 3, 3\}, 23\} & \{2, 3, 1, \{4, 3, 4\}, 21\} & \{2, 4, 1, \{5, 3, 1\}, 23\} \\ 0 & \{1, 5, 1, \{5, 3, 5\}, 12\} & 0 & \{1, 6, 0, \{2, 1, 5\}, 24\} \\ 0 & 0 & 0 & \{4, 7, 1, \{4, 5, 5\}, 13\} \end{pmatrix}$$

In most of the models in this book, there are multiple functions being applied to the lattice at each time step, so that the *increaseResources* step might be followed by a *walk* step (applied via the *GN* neighborhood function defined in the previous section). In such cases, this composition of functions can be given as a single anonymous function to Nest, as in the following case. (Since the walk rules have not been defined here, we do not evaluate this input.)

```
Nest[GN[walk,
        vonNeumann[increaseResources, #]]&, society, 10]
```

Similar to Nest and NestList are FixedPoint and FixedPointList, which can be used to stop the repeated application of a function when a specified condition is met, although we do not use them in this book. For a description of what they are and how they can be used, see the *Mathematica* book.

B.3.3 Fold and FoldList

We've seen how Nest and NestList can be used to iterate a function of one variable a fixed number of times. When a similar operation needs to be done with a function of two variables, Fold and FoldList can be used. We learn the syntax of FoldList from its usage message.

```
?FoldList

FoldList[f, x, {a, b, ... }] gives {x, f[x, a], f[f[
    x, a], b], ... }.
```

You can think of FoldList as iterating over its third argument. The following example uses Power, which is the *Mathematica* function for raising one number to the power of a second number. This code folds the Power function over the list $\{x, y, z\}$ starting with *a*.

```
FoldList[Power, a, {x, y, z}]
```

$$\{a, a^x, (a^x)^y, ((a^x)^y)^z\}$$

Just as Nest returns the last element of NestList, Fold returns the last element of FoldList.

```
Fold[Power, a, {x, y, z}]
```

$$((a^x)^y)^z$$

We use Nest, NestList, Fold, and FoldList as a more efficient substitute for recursive or procedural techniques in *Mathematica*.

B.3.4 Apply and Sequence

Up until now, we have used the term "apply a function" to casually refer to function application. So to "apply the function h to the list $\{1, 2, 3\}$" results in the expression $h[\{1, 2, 3\}]$. However, sometimes you want the function to *replace* the head of your argument so as to result in $h[1, 2, 3]$. To do this, you would use the built-in function Apply.

```
?Apply
```

> Apply[f, expr] or f @@ expr replaces the head of expr
>   by f. Apply[f, expr, levelspec] replaces heads in
>   parts of expr specified by levelspec.

A list of expressions, for example, could be added together by using Apply to replace the head List with the head Plus.

```
Apply[Plus, List[x, y, z]]
```

x + y + z

Summing lists can be useful, for instance, in calculating averages. Recall the matrix *society* created earlier in this appendix.

```
TraditionalForm[society]
```

$$\begin{pmatrix} 0 & 0 & 0 & \{3, 1, 1, \{2, 3, 4\}, 4\} \\ 0 & \{3, 2, 1, \{4, 3, 3\}, 3\} & \{2, 3, 1, \{4, 3, 4\}, 1\} & \{2, 4, 1, \{5, 3, 1\}, 3\} \\ 0 & \{1, 5, 1, \{5, 3, 5\}, 2\} & 0 & \{1, 6, 0, \{2, 1, 5\}, 4\} \\ 0 & 0 & 0 & \{4, 7, 1, \{4, 5, 5\}, 3\} \end{pmatrix}$$

We can extract a list consisting of each individual's resource level.

```
resources = Cases[society, {_, _, _, _, r_} :> r, {2}]
```

{4, 3, 1, 3, 2, 4, 3}

Then using Apply, we can calculate the average of this list by adding the elements and dividing by the number of elements.

```
Apply[Plus, resources] / Length[resources] // N
```

2.85714

Apply is not restricted to replacing a head of List, but will replace the head of any given expression.

```
Apply[h2, h[1, 2, 3]]
```

h2[*1, 2, 3*]

Related to Apply is Sequence, the *Mathematica* operator that denotes a comma-separated series of expressions. By using Apply to replace the head of an expression with Sequence, you can effectively splice the expression's arguments into another argument list.

```
h[1, 2, Apply[Sequence, {3, 4}], 5]
```

h[*1, 2, 3, 4, 5*]

Appendix C

Graphics and Efficiency

In this appendix, we discuss two kinds of functions that are useful for enhancing the practicality of a simulation. We demonstrate a wide variety of graphics commands, employing functions that are either built-in or accessible in the standard *Mathematica* graphics packages, and can be used to visualize the results of a simulation run. We then create an alternative function to Nest and NestList that uses less memory than these functions during a simulation run and also gives the user more flexibility in the storage and visualization of simulation data.

C.1 Graphics

The *Mathematica* language is extremely rich in graphics commands and options to alter graphical output. Here, we give a brief description of the graphics commands used in this book.

C.1.1 RasterArray

One of the most common ways to visualize a matrix is by displaying an array of colored squares. The position of the squares corresponds to the position of the individuals and empty sites in the lattice, and the color of the squares gives information about some aspect of the individuals. We can easily create an array like this by using the built-in function RasterArray, which takes a matrix of colors as its argument.

The colors can be given as shades of gray with the GrayLevel function. Other color functions include RGBColor, CMYKColor, and Hue. For this example, we begin with the same society used in Appendix B. Recall that empty sites are represented by a zero, and individuals have their behavior as the third argument to their list.

```
society = {{0, 0, 0, {3, 1, 1, {2, 3, 4}, 4}}, {0, {3, 2, 1,
{4, 3, 3}, 3}, {2, 3, 1, {4, 3, 4}, 1}, {2, 4, 1, {5, 3, 1},
3}}, {0, {1, 5, 1, {5, 3, 5}, 2}, 0, {1, 6, 0, {2, 1, 5},
4}}, {0, 0, 0, {4, 7, 1, {4, 5, 5}, 3}}};
TraditionalForm[society]
```

$$
\begin{pmatrix}
0 & 0 & 0 & \{3, 1, 1, \{2, 3, 4\}, 4\} \\
0 & \{3, 2, 1, \{4, 3, 3\}, 3\} & \{2, 3, 1, \{4, 3, 4\}, 1\} & \{2, 4, 1, \{5, 3, 1\}, 3\} \\
0 & \{1, 5, 1, \{5, 3, 5\}, 2\} & 0 & \{1, 6, 0, \{2, 1, 5\}, 4\} \\
0 & 0 & 0 & \{4, 7, 1, \{4, 5, 5\}, 3\}
\end{pmatrix}
$$

We observe that the individual named "6" is the only individual in the matrix that has a behavior of 0, and the rest have behavior 1. Using Map, we replace the empty sites with white (GrayLevel[1]) and individuals with either gray (GrayLevel[0.5]) or black (GrayLevel[0]) depending on whether they have behavior 1 or 0, respectively.

```
MatrixForm[
  Map[If[# === 0, GrayLevel[1], GrayLevel[#[[3]]/2]]&,
     society, {2}]]
```

$$
\begin{pmatrix}
GrayLevel[1] & GrayLevel[1] & GrayLevel[1] & GrayLevel[\frac{1}{2}] \\
GrayLevel[1] & GrayLevel[\frac{1}{2}] & GrayLevel[\frac{1}{2}] & GrayLevel[\frac{1}{2}] \\
GrayLevel[1] & GrayLevel[\frac{1}{2}] & GrayLevel[1] & GrayLevel[0] \\
GrayLevel[1] & GrayLevel[1] & GrayLevel[1] & GrayLevel[\frac{1}{2}]
\end{pmatrix}
$$

We can then display the grid as squares of colors using RasterArray. (The symbols Show, Graphics, Frame, and AspectRatio are described shortly.)

```
Show[Graphics[RasterArray[
  Map[If[# === 0, GrayLevel[1], GrayLevel[#[[3]]/2]]&,
     society, {2}]]],
  Frame → True, AspectRatio → Automatic];
```

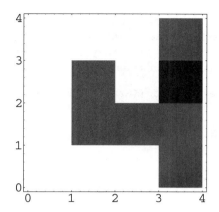

Notice that the first row of the matrix is drawn as the bottom row of the RasterArray graphic. In order to draw the graphic with the same orientation as that in which the matrix is displayed in MatrixForm, the order of the rows must be reversed. This is done by replacing *society* with Reverse[*society*].

```
Show[Graphics[RasterArray[
    Map[If[# === 0, GrayLevel[1], GrayLevel[#[[3]]/2]]&,
        Reverse[society], {2}]]],
    Frame → True, AspectRatio → Automatic];
```

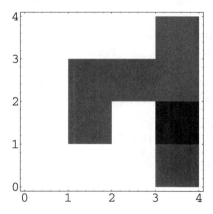

C.1.2 Graphics Primitives, Directives, and Options

As with RasterArray, the main syntax for displaying two-dimensional graphics objects is Show[Graphics[*listOfObjects*]]. This *listOfObjects* can contain graphics *primitives* and graphics *directives*. Graphics primitives should be thought of as building blocks from which other graphical objects are constructed. Examples of graphics primitives are Point, Line, and Polygon. The syntax for these primitives can be found in the online help, or by asking the kernel for their usage messages.

Graphics directives are used to alter the display of the graphics primitives that follow the directive in the *listOfObjects*. For instance, color is a graphics directive, so you can use an RGBColor or GrayLevel specification before a Polygon in order to draw a shaded polygon. Other directives include PointSize and Thickness.

```
Show[Graphics[{GrayLevel[0.7],
              Polygon[{{0, 0}, {0, 2}, {10, 1}}]}]];
```

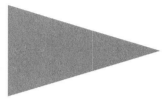

Another powerful way to alter the display of a graphic is to use *options*. A graphics option is either a Rule (→) or a RuleDelayed (:→) expression that specifies how a *Mathematica* graphics function should behave. Many built-in functions have options, but for the time being we focus on graphics options. A sequence of options is typically given to a function after its usual arguments. Thus, a sequence of options is inserted into the Show command as

```
Show[Graphics[listOfObjects], option1 → setting1,
          option2 → setting2, ..., optionk → settingk]
```

We've already seen an example of graphics options in the RasterArray command earlier in this appendix. Frame is an option that if set to True will draw a frame around the graphic. AspectRatio specifies the ratio of the height of a plot to its width, and a setting of Automatic chooses this ratio so that the same distance along both axes is drawn with the same length.

```
Show[Graphics[
  {GrayLevel[0.8], Polygon[{{0, 0}, {0, 2}, {10, 1}}]}],
  Frame → True,
  AspectRatio → Automatic,
  PlotLabel → "Gray Triangle"];
```

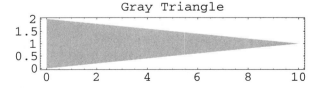

Other options, such as PlotLabel, and their default values can be listed by evaluating Options[Graphics]. You can change the default value of an option by using the SetOptions command.

Mathematica graphics are typically generated as PostScript, a text-based format which is a publishing and printing standard. One graphics option that is often used is

DisplayFunction, which controls how the graphic is displayed in *Mathematica*. By default, DisplayFunction is set to $DisplayFunction, which causes *Mathematica* to interpret and display PostScript as a picture like those you've seen in the preceding. A setting of DisplayFunction → Identity will entirely prevent the graphics from being displayed.

Later in this appendix, we show how to define a function that uses options to change its behavior.

C.1.3 ListPlot

When visualizing some value that changes over time, one function that is useful is ListPlot. ListPlot takes a list of y coordinates, and by default takes the x coordinate to be 1 for the first number, 2 for the second, and so on. It returns all these points in a graphic. Thus if you have the value of some quantity for every time step in a list *lis*, you can visualize this change over time by evaluating ListPlot[*lis*].

For example, in many of the models used in this book, an individual chooses one of four random directions to face during each time step. The following shows one such individual in isolation, where each point $\{x, y\}$ in the graph represents the direction y the individual faced at a given time step x.

```
r := Random[Integer, {1, 4}];
SeedRandom[0];
ListPlot[Table[r, {100}],
  Ticks → {{25, 50, 75, 100}, {2, 3, 4}},
  AxesOrigin → {0, 1}];
```

From this picture, it is difficult to get an idea of how an individual has changed direction from one time step to the next. In order to do this, we add the option setting PlotJoined → True to the call to ListPlot, which tells *Mathematica* to draw a line segment between consecutive points.

```
r := Random[Integer, {1, 4}];
SeedRandom[0];
ListPlot[Table[r, {100}],
```

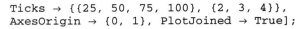

```
Ticks → {{25, 50, 75, 100}, {2, 3, 4}},
AxesOrigin → {0, 1}, PlotJoined → True];
```

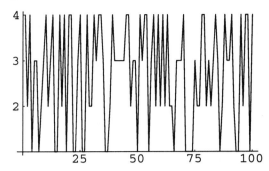

If you give ListPlot a list of ordered pairs instead of a list, it plots each ordered pair {x, y} as a point with the given x and y coordinates.

C.1.4 Graphics Packages

In addition to built-in graphics functions, there are many more functions defined in the standard packages that ship with *Mathematica*. In order to automatically load the graphics packages, we load the Graphics *master package*, which the following input does.

```
<<Graphics`
```

This loads the names of the package-defined graphics functions into the kernel, so that when they are used, the associated package is loaded. Without first loading the master package, the examples in this section will not work.

MultipleListPlot

We have already seen how to plot the value of some quantity as a function of time by using ListPlot. If we want to simultaneously plot the values of several quantities over the same period of time, in order to see how each changes with respect to the others, MultipleListPlot can be used. Giving MultipleListPlot a sequence of lists, it plots each list as ListPlot would, using different symbols instead of points for the different data sets. For instance, suppose we know that the resource levels for two individuals satisfy the following, for $n = 1$ and $n = 2$.

```
resources[n_] := (SeedRandom[n];
          NestList[(Random[Real, {-1, 1}] + #&), 0, 300])
```

We can plot the resource levels of these two individuals over time on the same set of axes using MultipleListPlot, which uses diamonds and stars to represent the two different data sets.

```
MultipleListPlot[resources[1], resources[2],
   AxesLabel → {"time", "values"}];
```

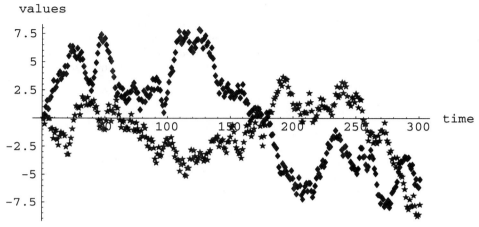

Other options to MultipleListPlot allow us to connect the data points with lines that allow us to dfferentiate between them, instead of using different symbols for the points.

```
MultipleListPlot[resources[1], resources[2],
   AxesLabel → {"time", "values"},
   SymbolShape → None,
   PlotJoined → True];
```

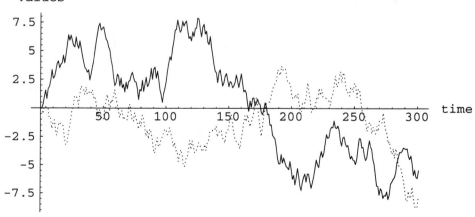

Notice that these two plots don't allow us to know which line refers to which set of data points. To do this, we can add a legend to this plot by using another package: PlotLegend.

PlotLegend

PlotLegend is an option that allows us to add a legend to plots that use different styles for different data. It gives a graphical representation of which style goes with which data set. For instance, suppose resources[1] is the resource list of an individual we have labeled as *Type a*, and resources[2] corresponds to an individual that is *Type b*. We can add a PlotLegend to the preceding graphic that illustrates that correspondence.

```
MultipleListPlot[resources[1], resources[2],
   AxesLabel → {"time", "values"},
   SymbolShape → None,
   PlotJoined → True,
   PlotLegend → {"Type a", "Type b"}];
```

There are other options that affect where a legend is drawn with respect to the rest of the graphic (LegendPosition), its size (LegendSize), whether it is drawn as a shadow box (LegendShadow), and many others.

You can get more information about these options by asking for their usage messages, or by looking in the online help for the Graphics`Legend` package.

```
MultipleListPlot[resources[1], resources[2],
   AxesLabel → {"time", "values"},
   SymbolShape → None,
   PlotJoined → True,
   PlotLegend → {"Type a", "Type b"},
   LegendOrientation → Horizontal,
   LegendSize → {1, 0.3},
   LegendPosition → {-1, -0.8},
   LegendShadow → {-0.05, -0.07}];
```

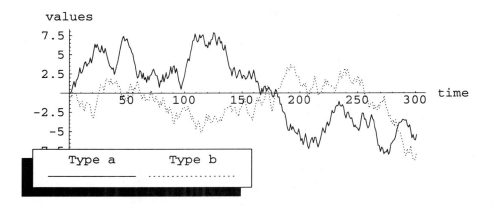

Bar Chart

Another graphics function that is used throughout this book is BarChart. As with MultipleListPlot, BarChart takes a list or a sequence of lists, and produces a graphic displaying those data sets.

For example, suppose that in some model, individuals can switch from one group to another, and we are interested in the number of individuals in a particular group as a function of time. Say we have already extracted the number of individuals in each group for every time step, and have the results in two lists: *group1* and *group2*.

```
group1 = {10,11,10,11,10,9,8,9,10,9,8,7,6,5,6,7,8,9,8,7,8};
group2 = {10,9,10,9,10,11,12,11,10,11,12,13,14,15,
                                   14,13,12,11,12,13,12};
```

Plotting these data as a bar chart is straightforward.

```
BarChart[group1, group2, AxesLabel → {"time", "count"},
  Ticks → {Range[0, 20, 2], Range[0, 14, 2]}];
```

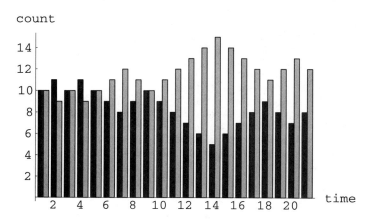

In the Graphics`Graphics` package, a number of variations on the standard BarChart are defined. For example, even though individuals are switching between group 1 and group 2 in our example, the total population remains the same. Thus, it makes sense to plot the size of each group as a percentage of the total population. We can do this with a PercentileBarChart.

```
PercentileBarChart[group1, group2,
  AxesLabel → {"time", "density"},
  BarStyle → {GrayLevel[0.4], GrayLevel[0.8]}];
```

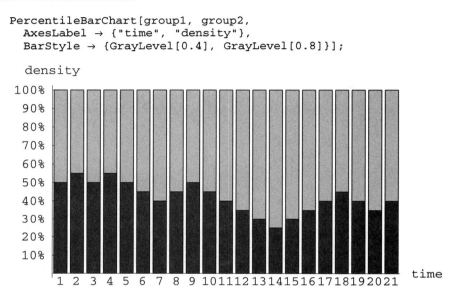

C.1.5 GraphicsArray

The GraphicsArray function is used when you want to display a table of complete graphics. The syntax is Show[GraphicsArray[*matrixOfGraphics*]]. Here is a simple example, combining two of the graphics we've already created. Notice that we use the DisplayFunction option to suppress the graphics from being displayed until the very end.

```
g1 = MultipleListPlot[resources[1], resources[2],
  AxesLabel → {"time", "values"},
  SymbolShape → None,
  PlotJoined → True,
  PlotLegend → {"Type a", "Type b"},
  LegendOrientation → Horizontal,
  LegendSize → {1, 0.3},
  LegendPosition → {-1, -0.8},
  LegendShadow → {-0.05, -0.07},
  DisplayFunction → Identity];
g2 = PercentileBarChart[group1, group2,
  PlotLabel → "Density vs. Time",
  Ticks → None,
  BarStyle → {GrayLevel[0.4], GrayLevel[0.8]},
  DisplayFunction → Identity];
```

```
Show[GraphicsArray[{g1, g2}],
    DisplayFunction → $DisplayFunction];
```

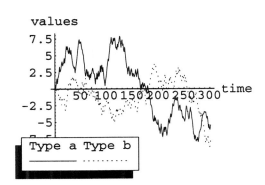

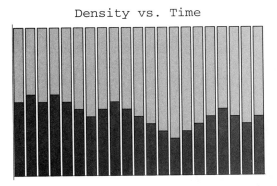

C.1.6 Animation

Mathematica allows us to generate a sequence of graphics and view it as a movie or animation. The most common way to animate graphics is simply to generate a sequence of graphics using Map or Table. When all the graphics have been generated, double-click on any one of them. *Mathematica* will then show them, one by one, in a flip-book style of animation. Unfortunately, having all these graphics collected in the same notebook can be very memory and/or disk intensive.

With the new technology available in the *Mathematica* version 3 front end, we can generate these graphics one at a time, and have each new graphic displace the old one. In this way, we can generate the flip-book animation "on the fly," without having to save these graphics to disk at all. The next section describes the front end programming techniques necessary to accomplish this.

C.2 Animations and Front End Programming

On all platforms, the *Mathematica* version 3 front end supports a number of commands that enable a kernel program to do almost anything that a user would do in a typical session. From creating notebooks and cells to cutting and pasting to printing, the front end can do all these things under control of the kernel. This section gives a brief introduction to front end programming. For more information, refer to *Mathematica*'s online documentation.

One of the major changes in version 3 of *Mathematica* is that notebooks are simply expressions. As a result, the kernel can be used to create and manipulate notebook expressions. In this section, we describe how to create and manipulate graphics cells from the kernel, allowing us to animate graphics on the fly. We generate them by creating a new notebook and then repeatedly replacing the contents of that notebook with the next graphics cell.

Creating a New Notebook

The animation will run in a new window in the front end. To create a new notebook window, we can use the built-in function NotebookCreate. Evaluating the following input cell causes a new untitled notebook to open in the front end. We store the result in a variable (*nb*) so we can access that same notebook with subsequent kernel commands.

```
nb = NotebookCreate[];
```

Writing Information to a Notebook

There are a number of notebook functions that allow us to put information into a notebook: NotebookPut, NotebookApply, NotebookWrite, CellPrint, and so on. We use NotebookWrite, which typically takes two arguments. The first argument indicates to which open notebook you are writing, and the second argument indicates the expression to be written into that notebook. The expression is typically a string, a Cell, or a list of Cell expressions. We discuss Cell expressions shortly. This example writes a string corresponding to a *Mathematica* function call into the notebook.

```
NotebookWrite[nb,
        "ListPlot[resources[1],PlotJoined → True]"];
```

NotebookWrite can also be used to replace any selected cells with the data to be written. Thus, you can replace the contents of the notebook *nb* by first selecting the entire notebook, which the following code does.

```
SelectionMove[nb, All, Notebook]
```

After this selection, use NotebookWrite to write the new data in place of the old.

```
NotebookWrite[nb,
        "MultipleListPlot[resources[1], resources[2]]"];
```

Thus, we can get a simple flip-book animation of an input cell by replacing old input cells with new ones. Running this in the online version of the book, notice that there is never more than one cell in the animation notebook at a time.

```
nb2 = NotebookCreate[];
Table[SelectionMove[nb2, All, Notebook];
      NotebookWrite[nb2, ToString[i]], {i, 1, 10}];
```

Graphics Cells

In order to use the preceding technique to animate graphics cells, we must first learn the syntax of such cells. These cells are *Mathematica* expressions, just as notebooks are. A typical graphics cell has the following syntax.

```
Cell[GraphicsData["PostScript", postscriptText], "Graphics"]
```

The variable *postscriptText* refers to the PostScript that generates the picture, which is really just a long string. One way to access the PostScript text behind graphics is by changing the setting of the DisplayFunction option to the function DisplayString, which has the effect of returning just the PostScript string for the graphic. For example, here we generate the PostScript string corresponding to the RasterArray earlier in this appendix.

```
society = {{0, 0, 0, {3, 1, 1, {2, 3, 4}, 4}}, {0, {3, 2, 1,
{4, 3, 3}, 3}, {2, 3, 1, {4, 3, 4}, 1}, {2, 4, 1, {5, 3, 1},
3}}, {0, {1, 5, 1, {5, 3, 5}, 2}, 0, {1, 6, 0, {2, 1, 5},
4}}, {0, 0, 0, {4, 7, 1, {4, 5, 5}, 3}}};
str = Show[Graphics[RasterArray[
  Map[If[# === 0, GrayLevel[1], GrayLevel[#[[3]]/2]]&,
    society, {2}]]],
  Frame → True,
  AspectRatio → Automatic,
  DisplayFunction → DisplayString];
```

Now that we have this string, we can plug it into the preceding format for graphics cells and write it into the notebook, replacing any cells that were in the notebook already.

```
nb = NotebookCreate[];
SelectionMove[nb, All, Notebook];
NotebookWrite[nb,
    Cell[GraphicsData["PostScript", str], "Graphics"]];
```

Animation

Iterating this process, we can generate animations as graphics are generated, keeping only the last graphic that has been generated in this way. In the next section, we discuss how to add this capability into your simulations, so that your societies can be shown changing as the changes are taking place during the simulation run.

C.3 Simulations in Limited Memory

In this section, we introduce a custom function which can replace the Nest and NestList functions that allows us to focus on simulation results without a lot of worrying about how much memory the simulation will use. It also allows us to save information at various times during the simulation, as well as to view graphics during the simulation run. Most important, it does all this with as little change to the simulation program as possible.

C.3.1 NestListEffects

Throughout the book, we use NestList to apply a model's rule sets iteratively. As a result, it returns a list of every lattice configuration of the simulation run. If we chose to use Nest instead of NestList, it would only return the last step in the simulation. This results in considerable savings in memory consumption.

To illustrate this, we use one of the programs from Chapter 5, the *neighborhood* program, with both Nest and NestList. We call the version that uses NestList, *neighborhoodNL*, and the program that uses Nest, *neighborhoodN*.

```
neighborhoodNL[n_, p_, v_, w_, t_] :=
Module[  (* program goes here *)
  NestList[GN[walk, Moore[movestay, #]]&, society, t]]

neighborhoodN[n_, p_, v_, w_, t_] :=
Module[  (* program goes here *)
  Nest[GN[walk, Moore[movestay, #]]&, society, t]]
```

We can check how memory-intensive each of these functions is by using the built-in functions MemoryInUse and MaxMemoryUsed.

```
m1 = {MemoryInUse[], MaxMemoryUsed[]};
len = Length[neighborhoodNL[20, 0.60, 1, 2, 1000]];
m2 = {MemoryInUse[], MaxMemoryUsed[]};
{len, m2 - m1}

{1001, {1688, 7798392}}
```

We see that during the simulation, almost 8Mb of memory is consumed, although *Mathematica* frees most of the memory when the simulation has completed. Compare this with the results from using Nest in a fresh kernel session.

```
m1 = {MemoryInUse[], MaxMemoryUsed[]};
len = Length[{neighborhoodN[20, 0.60, 1, 2, 1000]}];
m2 = {MemoryInUse[], MaxMemoryUsed[]};
{len, m2 - m1}

{1, {1672, 81120}}
```

So Nest uses slightly more than 81K or 0.08Mb of memory, a savings factor of nearly 100 over NestList.

Besides using Nest and NestList, there is a third possibility: NestListEffects. This function, defined at the end of this appendix, enables us to do two things: it allows us to specify what lattice configurations should be returned with greater flexibility than NestList or Nest; and it allows us to generate side effects during the course of the function application, such as printing the memory that is in use, or generating a graphic corresponding to the current lattice configuration. These two features are controlled through options to NestListEffects named ReturnValues and SideEffects, respectively.

Instead of returning every element of NestList, you can return every kth element by setting the ReturnValues option to {(#1 &), k}. In order to return every kth element after applying the function g to those elements, set ReturnValues to {g, k}. The function g actually takes two arguments, where the second argument is the position that the current element would have in the output of NestList. To illustrate this, we use the following simple example.

```
NestList[# + 2 &, 0, 10]

{0, 2, 4, 6, 8, 10, 12, 14, 16, 18, 20}
```

Setting ReturnValues to {(#1 &), 3} returns every third element of the corresponding NestList function, starting with the second element.

```
NestListEffects[# + 2 &, 0, 10,
  ReturnValues → {(#1 &), 3}]

{2, 8, 14, 20}
```

The SideEffects option is set to a list of ordered pairs, each of which functions in the same way as the setting of ReturnValues does. Each ordered pair {g_i, k_i} consists of a function of two arguments and an integer that specifies how often to apply the function. For example, here's a side effect that displays the step number during every other step. The function *effect1* uses Print to display each of these step numbers on a line by itself, and then the final result is returned.

```
effect1[x_, counter_] := Print[counter];
NestListEffects[# + 2 &, 0, 10,
  ReturnValues → {(#1 &), 3},
  SideEffects → {{effect1, 2}}]

0

2

4

6

8

{2, 8, 14, 20}
```

Another possible side effect would be to display a graphic representing the lattice configuration to a notebook window, replacing the contents of that window in the process. In this way, we can run an animation during the simulation run itself without having to save every graphic first. In the following, we explore the code that does this.

Adding Effects to a Simulation

After changing NestList to NestListEffects in the simulation code, there is one more thing we need to do: allow the user to pass options to NestListEffects. We do this by changing the left-hand side of the simulation program's definition so that the last optional arguments are saved in a sequence named *opts*. For example, if we call this new program *neighborhoodNLE*, then the new left-hand side of the program should look like the following. Notice that *opts* will match zero or more arguments (options) because of the triple blank (___).

```
neighborhoodNLE[n_, p_, v_, w_, t_, opts___]
```

Then these options can be passed directly to NestListEffects by refering to *opts*.

```
NestListEffects[
        GN[walk, Moore[movestay, #]]&, society, t, opts]
```

Now this new function, *neighborhoodNLE*, is ready to accept ReturnValues and SideEffects options.

```
neighborhoodNLE[n_, p_, v_, w_, t_, opts___] :=
Module[  (* program goes here *)
   NestListEffects[
        GN[walk, Moore[movestay, #]]&, society, t, opts]]
```

Kernel Memory Savings

Replacing NestList with NestListEffects in this way achieves almost as much memory savings as Nest afforded.

```
m1 = {MemoryInUse[], MaxMemoryUsed[]};
len = Length[neighborhoodNLE[20, 0.60, 1, 2, 1000,
                      ReturnValues → {(#1 &), 99}]];
m2 = {MemoryInUse[], MaxMemoryUsed[]};
{len, m2 - m1}

{11, {2672, 197280}}
```

This uses just less than 200K, which is slightly more than Nest, but still drastically less than NestList, and with much more flexibility and control of the values returned from the run.

Another possible setting for ReturnValues would not return the entire lattice, but some value or values extracted from it. For example, the number of empty sites should remain constant for each time step in this model, and we can check that that is happening by applying Count[#, 0, {2}]& to the lattice every so often. The following example does that, and also uses an effect named *printCounter*, defined at the end of this appendix, to display the simulation's progress.

```
neighborhoodNLE[20, 0.60, 1, 2, 100,
   ReturnValues → {Count[#, 0, {2}]&, 20},
   SideEffects → {{printCounter, 10}}]
```

0 10 20 30 40 50 60 70 80 90

{153, 153, 153, 153, 153}

Front End Memory Savings

Notice that this technique not only allows the kernel to save memory, but also the front end. For example, this gives us the ability to generate a graphic during each time step, and replace it with a new graphic during the next time step. In this way, we see the animation as it develops, without having to save possibly thousands of memory-hungry graphics.

```
displayLattice[nb_, lat_, counter_] :=
Module[{str},
   SelectionMove[nb, All, Notebook];
   str = Show[Graphics[RasterArray[
      lat /. {0 → RGBColor[0.7, 0.7, 0.7],
            {_, {1}} → RGBColor[0, 1, 0],
            {_, {2}} → RGBColor[0, 0, 1]}]],
      PlotLabel → ("lattice after time step " <>
                  ToString[counter]),
      AspectRatio → Automatic,
      DisplayFunction → DisplayString];
   NotebookWrite[nb,
      Cell[GraphicsData["PostScript", str], "Graphics"]];]

nb = NotebookCreate[];
neighborhoodNLE[20, 0.60, 1, 2, 100,
   SideEffects → {{displayLattice[nb, #1, #2]&, 1}}];
```

Using multiple side effects, we can easily set up more than one animation running at the same time, each being updated at different intervals, as well as having data written into files or other actions occurring simultaneously.

C.3.2 Options and Genetic Algorithms

In addition to ReturnValues and SideEffects, one can add other options to the simulation programs. For instance, in all the examples in this book we have the simulations beginning with random lattice configurations. In some applications, such as applying genetic algorithms, it may be necessary to define the initial configuration by hand or as the result of some other calculation.

One way to do this is to add an option to the simulation program that will enable the user to specify an initial configuration instead of always beginning with a random configuration. Say we want to call such an option InitialConfiguration. The default value for this option should be Automatic, meaning that a random configuration will be used. If the setting is something other than Automatic, the setting itself should be used as the starting lattice for the simulation.

To do this, we first define the default value. For the program name, we use yet another variant of the neighborhood model, *neighborhoodOpts*, which stands for "neighborhood with options."

```
Options[neighborhoodOpts] =
                {InitialConfiguration → Automatic};
```

Next, we need to add the code that reads the value of the option and then acts accordingly. Since options are expressions with head Rule, we can extract the setting for InitialConfiguration using ReplaceAll (/.) and save it in a new local variable called *initConfig*. Then, if this is set to Automatic, the code generates a random lattice configuration; otherwise, it uses the setting that the user specifies.

```
neighborhoodOpts[n_, p_, v_, w_, t_, opts___] :=
Module[{RND, initConfig, society},
   RND := Random[Integer, {1, 4}];
   initConfig = InitialConfiguration /.
                {opts} /. Options[neighborhoodOpts];
   society = If[initConfig === Automatic,
     Table[Floor[p + Random[]], {n}, {n}] /.
         1 :→ {RND, Table[Random[Integer, {1, w}], {v}]},
     initConfig];

(* program goes here *)

NestList[GN[walk, Moore[movestay, #]]&, society, t]]
```

This enables one to pass in a configuration that was, for example, output from some previous simulation run, perhaps after altering it in some way. This is exactly how one might add genetic algorithms to a simulation. For example, here we run the same simulation in two ways. The first method is to run the entire simulation for 100 time steps. The second method is to run the simulation for 20 time steps, pass the resulting lattice configuration through as the initial configuration for the next 20 time steps, and so on. As you might expect, both methods result in the same final configuration.

```
SeedRandom[0];
config1 = Last[neighborhoodNL[20, 0.60, 1, 2, 100]];

SeedRandom[0];
config2 = Nest[Last[neighborhoodOpts[20, 0.60, 1, 2, 20,
              InitialConfiguration → #]]&, Automatic, 5];

config1 === config2

True
```

C.4 Programs in the Appendix

C.4.1 NestListEffects, ReturnValues, and SideEffects

```
NestListEffects::usage = "NestListEffects[f, expr, n,
ReturnValues → {g, k}] returns a list of every kth element
of NestList[f, expr, n] after the application of g.
NestListEffects[f, expr, n, SideEffects → {{g1, k1}, {g2,
k2},...}] applies gi after every ki steps, but does not
effect return values.";

ReturnValues::usage = "ReturnValues is an option to
NestListEffects that determines what is returned and how
often.";

SideEffects::usage = "SideEffects is an option to
NestListEffects that applies functions without effecting
return values.";

Options[NestListEffects] =
                {ReturnValues → {#&, 1}, SideEffects→{}};

NestListEffects[f_, expr_, n_, opts___] :=
If[({ReturnValues, SideEffects} /. {opts} /.
        Options[NestListEffects]) === {{#&, 1}, {}},
  NestList[f, expr, n],
  (* else *)
  Module[{c = 0, lis = {}, side, ret},
    {ret, side} = {ReturnValues, SideEffects} /.
                    {opts} /. Options[NestListEffects];
    If[side === {}, side = {{Null, Infinity}}];
    Nest[(
      If[Mod[c, ret[[-1]]] === 0,
        lis = Join[lis, {First[ret][#, c]}]];
      Table[If[Mod[c, side[[i, -1]]] === 0,
              side[[i, 1]][#, c]],
        {i,Length[side]}];
      c++; #)&[f[#]]&, expr, n];
    lis]
];
```

C.4.2 neighborhoodNL

```
neighborhoodNL[n_, p_, v_, w_, t_] :=
Module[{walk, movestay, society, RND, Moore, GN},
  RND := Random[Integer, {1, 4}];
  society = Table[Floor[p + Random[]], {n}, {n}] /.
        1 :→ {RND, Table[Random[Integer, {1, w}], {v}]};
  movestay[0, __] := 0;
  movestay[{a_, b_}, res__] :=
    {a * Round[1 - Count[Map[Count[b - #[[2]], 0]&,
```

```
              {res} /. 0 → {0, 0}], _?(# >= v / 2 &)] / 8.], b};
   walk[{1,a_},0,_,_,_,{4,_},_,_,_,_,_,_,_] := {RND,a};
   walk[{1,a_},0,_,_,_,_,_,_,{2,_},_,_,_,_] := {RND,a};
   walk[{1,a_},0,_,_,_,_,_,_,_,{3,_},_,_,_] := {RND,a};
   walk[{1,a_},0,_,_,_,_,_,_,_,_,_,_,_] := 0;
   walk[{2,a_},_,0,_,_,{3,_},_,_,_,_,_,_,_] := {RND,a};
   walk[{2,a_},_,0,_,_,_,{1,_},_,_,_,_,_,_] := {RND,a};
   walk[{2,a_},_,0,_,_,_,_,_,{4,_},_,_,_] := {RND,a};
   walk[{2,a_},_,0,_,_,_,_,_,_,_,_,_,_] := 0;
   walk[{3,a_},_,_,0,_,_,{4,_},_,_,_,_,_,_] := {RND,a};
   walk[{3,a_},_,_,0,_,_,_,{2,_},_,_,_,_,_] := {RND,a};
   walk[{3,a_},_,_,0,_,_,_,_,_,_,{1,_},_] := {RND,a};
   walk[{3,a_},_,_,0,_,_,_,_,_,_,_,_,_] := 0;
   walk[{4,a_},_,_,_,0,_,_,{1,_},_,_,_,_,_] := {RND,a};
   walk[{4,a_},_,_,_,0,_,_,_,{3,_},_,_,_,_] := {RND,a};
   walk[{4,a_},_,_,_,0,_,_,_,_,_,_,{2,_}] := {RND,a};
   walk[{4,a_},_,_,_,0,_,_,_,_,_,_,_,_] := 0;
   walk[{x_,a_},_,_,_,_,_,_,_,_,_,_,_,_] := {RND,a};
   walk[0,{3,_},{4,_},_,_,_,_,_,_,_,_,_,_] := 0;
   walk[0,{3,_},_,{1,_},_,_,_,_,_,_,_,_,_] := 0;
   walk[0,{3,_},_,_,{2,_},_,_,_,_,_,_,_,_] := 0;
   walk[0,_,{4,_},{1,_},_,_,_,_,_,_,_,_,_] := 0;
   walk[0,_,{4,_},_,{2,_},_,_,_,_,_,_,_,_] := 0;
   walk[0,_,_,{1,_},{2,_},_,_,_,_,_,_,_,_] := 0;
   walk[0,{3,a_},_,_,_,_,_,_,_,_,_,_,_] := {RND,a};
   walk[0,_,{4,a_},_,_,_,_,_,_,_,_,_,_] := {RND,a};
   walk[0,_,_,{1,a_},_,_,_,_,_,_,_,_,_] := {RND,a};
   walk[0,_,_,_,{2,a_},_,_,_,_,_,_,_,_] := {RND,a};
   walk[0,_,_,_,_,_,_,_,_,_,_,_,_] := 0;
   Moore[func_, lat_] :=
     MapThread[func, Map[RotateRight[lat, #]&,
             {{0, 0}, {1, 0}, {0, -1}, {-1, 0}, {0, 1},
              {1, -1}, {-1, -1}, {-1, 1}, {1, 1}}], 2];
   GN[func_, lat_] :=
     MapThread[func, Map[RotateRight[lat, #]&,
             {{0, 0}, {1, 0}, {0, -1}, {-1, 0}, {0, 1},
              {1, -1}, {-1, -1}, {-1, 1}, {1, 1}, {2, 0},
              {0, -2}, {-2, 0}, {0, 2}}], 2];
   NestList[GN[walk, Moore[movestay, #]]&, society, t]]
```

C.4.3 neighborhoodN

```
neighborhoodN[n_, p_, v_, w_, t_] :=
Module[{walk, movestay, society, RND, Moore, GN},
   RND := Random[Integer, {1, 4}];
   society = Table[Floor[p + Random[]], {n}, {n}] /.
         1 :> {RND, Table[Random[Integer, {1, w}], {v}]};
   movestay[0, __] := 0;
   movestay[{a_, b_}, res__] :=
      {a * Round[1 - Count[Map[Count[b - #[[2]], 0]&,
         {res} /. 0 → {0, 0}], _?(# >= v / 2 &)] / 8.], b};
   walk[{1,a_},0,_,_,_,{4,_},_,_,_,_,_,_,_] := {RND,a};
   walk[{1,a_},0,_,_,_,_,_,_,{2,_},_,_,_,_] := {RND,a};
   walk[{1,a_},0,_,_,_,_,_,_,_,{3,_},_,_,_] := {RND,a};
```

```
walk[{1,a_},0,_,_,_,_,_,_,_,_,_,_] := 0;
walk[{2,a_},_,0,_,_,{3,_},_,_,_,_,_,_] := {RND,a};
walk[{2,a_},_,0,_,_,_,{1,_},_,_,_,_,_] := {RND,a};
walk[{2,a_},_,0,_,_,_,_,_,{4,_},_,_] := {RND,a};
walk[{2,a_},_,0,_,_,_,_,_,_,_,_] := 0;
walk[{3,a_},_,_,0,_,_,{4,_},_,_,_,_,_] := {RND,a};
walk[{3,a_},_,_,0,_,_,_,{2,_},_,_,_,_] := {RND,a};
walk[{3,a_},_,_,0,_,_,_,_,_,{1,_},_] := {RND,a};
walk[{3,a_},_,_,0,_,_,_,_,_,_,_] := 0;
walk[{4,a_},_,_,_,0,_,_,{1,_},_,_,_,_] := {RND,a};
walk[{4,a_},_,_,_,0,_,_,_,{3,_},_,_] := {RND,a};
walk[{4,a_},_,_,_,0,_,_,_,_,_,{2,_}] := {RND,a};
walk[{4,a_},_,_,_,0,_,_,_,_,_,_] := 0;
walk[{x_,a_},_,_,_,_,_,_,_,_,_,_] := {RND,a};
walk[0,{3,_},{4,_},_,_,_,_,_,_,_,_] := 0;
walk[0,{3,_},_,{1,_},_,_,_,_,_,_,_] := 0;
walk[0,{3,_},_,_,{2,_},_,_,_,_,_,_] := 0;
walk[0,_,{4,_},{1,_},_,_,_,_,_,_,_] := 0;
walk[0,_,{4,_},_,{2,_},_,_,_,_,_,_] := 0;
walk[0,_,_,{1,_},{2,_},_,_,_,_,_,_] := 0;
walk[0,{3,a_},_,_,_,_,_,_,_,_,_] := {RND,a};
walk[0,_,{4,a_},_,_,_,_,_,_,_,_] := {RND,a};
walk[0,_,_,{1,a_},_,_,_,_,_,_,_] := {RND,a};
walk[0,_,_,_,{2,a_},_,_,_,_,_,_] := {RND,a};
walk[0,_,_,_,_,_,_,_,_,_] := 0;
Moore[func_, lat_] :=
  MapThread[func, Map[RotateRight[lat, #]&,
       {{0, 0}, {1, 0}, {0, -1}, {-1, 0}, {0, 1},
        {1, -1}, {-1, -1}, {-1, 1}, {1, 1}}], 2];
GN[func_, lat_] :=
  MapThread[func, Map[RotateRight[lat, #]&,
       {{0, 0}, {1, 0}, {0, -1}, {-1, 0}, {0, 1},
        {1, -1}, {-1, -1}, {-1, 1}, {1, 1}, {2, 0},
        {0, -2}, {-2, 0}, {0, 2}}], 2];
Nest[GN[walk, Moore[movestay, #]]&, society, t]]
```

C.4.4 neighborhoodNLE

```
neighborhoodNLE[n_, p_, v_, w_, t_, opts___] :=
Module[{walk, movestay, society, RND, Moore, GN},
  RND := Random[Integer, {1, 4}];
  society = Table[Floor[p + Random[]], {n}, {n}] /.
       1 :> {RND, Table[Random[Integer, {1, w}], {v}]};
  movestay[0, __] := 0;
  movestay[{a_, b_}, res__] :=
     {a * Round[1 - Count[Map[Count[b - #[[2]], 0]&,
       {res} /. 0 -> {0, 0}], _?(# >= v / 2 &)] / 8.], b};
  walk[{1,a_},0,_,_,_,{4,_},_,_,_,_,_,_] := {RND,a};
  walk[{1,a_},0,_,_,_,_,_,{2,_},_,_,_,_] := {RND,a};
  walk[{1,a_},0,_,_,_,_,_,_,_,{3,_},_,_] := {RND,a};
  walk[{1,a_},0,_,_,_,_,_,_,_,_,_] := 0;
  walk[{2,a_},_,0,_,_,{3,_},_,_,_,_,_,_] := {RND,a};
  walk[{2,a_},_,0,_,_,_,{1,_},_,_,_,_,_] := {RND,a};
  walk[{2,a_},_,0,_,_,_,_,_,{4,_},_,_] := {RND,a};
```

```
walk[{2,a_},_,0,_,_,_,_,_,_,_,_,_,_] := 0;
walk[{3,a_},_,_,0,_,_,{4,_},_,_,_,_,_,_] := {RND,a};
walk[{3,a_},_,_,0,_,_,_,{2,_},_,_,_,_] := {RND,a};
walk[{3,a_},_,_,0,_,_,_,_,_,{1,_},_] := {RND,a};
walk[{3,a_},_,_,0,_,_,_,_,_,_,_] := 0;
walk[{4,a_},_,_,_,0,_,_,{1,_},_,_,_,_] := {RND,a};
walk[{4,a_},_,_,_,0,_,_,_,{3,_},_,_,_] := {RND,a};
walk[{4,a_},_,_,_,0,_,_,_,_,_,_,{2,_}] := {RND,a};
walk[{4,a_},_,_,_,0,_,_,_,_,_,_,_] := 0;
walk[{x_,a_},_,_,_,_,_,_,_,_,_,_,_,_] := {RND,a};
walk[0,{3,_},{4,_},_,_,_,_,_,_,_,_,_] := 0;
walk[0,{3,_},_,{1,_},_,_,_,_,_,_,_,_] := 0;
walk[0,{3,_},_,_,{2,_},_,_,_,_,_,_,_] := 0;
walk[0,_,{4,_},{1,_},_,_,_,_,_,_,_,_] := 0;
walk[0,_,{4,_},_,{2,_},_,_,_,_,_,_,_] := 0;
walk[0,_,_,{1,_},{2,_},_,_,_,_,_,_,_] := 0;
walk[0,{3,a_},_,_,_,_,_,_,_,_,_,_] := {RND,a};
walk[0,_,{4,a_},_,_,_,_,_,_,_,_,_] := {RND,a};
walk[0,_,_,{1,a_},_,_,_,_,_,_,_,_] := {RND,a};
walk[0,_,_,_,{2,a_},_,_,_,_,_,_,_] := {RND,a};
walk[0,_,_,_,_,_,_,_,_,_,_,_] := 0;
Moore[func_, lat_] :=
  MapThread[func, Map[RotateRight[lat, #]&,
         {{0, 0}, {1, 0}, {0, -1}, {-1, 0}, {0, 1},
          {1, -1}, {-1, -1}, {-1, 1}, {1, 1}}], 2];
GN[func_, lat_] :=
  MapThread[func, Map[RotateRight[lat, #]&,
          {{0, 0}, {1, 0}, {0, -1}, {-1, 0}, {0, 1},
           {1, -1}, {-1, -1}, {-1, 1}, {1, 1}, {2, 0},
           {0, -2}, {-2, 0}, {0, 2}}], 2];
NestListEffects[
     GN[walk, Moore[movestay, #]]&, society, t, opts]]
```

C.4.5 neighborhoodOpts

```
neighborhoodOpts[n_, p_, v_, w_, t_, opts___] :=
Module[{walk, movestay, society, RND,
                                Moore, GN, initConfig},
  RND := Random[Integer, {1, 4}];
  initConfig = InitialConfiguration /.
                  {opts} /. Options[neighborhoodOpts];
  society = If[initConfig === Automatic,
    Table[Floor[p + Random[]], {n}, {n}] /.
       1 :> {RND, Table[Random[Integer, {1, w}], {v}]},
    initConfig];
  movestay[0, __] := 0;
  movestay[{a_, b_}, res__] :=
    {a * Round[1 - Count[Map[Count[b - #[[2]], 0]&,
     {res} /. 0 -> {0, 0}], _?(# >= v / 2 &)] / 8.], b};
  walk[{1,a_},0,_,_,_,{4,_},_,_,_,_,_] := {RND,a};
  walk[{1,a_},0,_,_,_,_,{2,_},_,_,_,_] := {RND,a};
  walk[{1,a_},0,_,_,_,_,_,{3,_},_,_,_] := {RND,a};
  walk[{1,a_},0,_,_,_,_,_,_,_,_,_] := 0;
  walk[{2,a_},_,0,_,_,{3,_},_,_,_,_,_,_] := {RND,a};
```

```
walk[{2,a_},_,0,_,_,_,{1,_},_,_,_,_,_,_] := {RND,a};
walk[{2,a_},_,0,_,_,_,_,_,_,_,{4,_},_,_] := {RND,a};
walk[{2,a_},_,0,_,_,_,_,_,_,_,_,_,_] := 0;
walk[{3,a_},_,_,0,_,_,_,{4,_},_,_,_,_,_,_] := {RND,a};
walk[{3,a_},_,_,0,_,_,_,{2,_},_,_,_,_,_] := {RND,a};
walk[{3,a_},_,_,0,_,_,_,_,_,_,_,{1,_},_] := {RND,a};
walk[{3,a_},_,_,0,_,_,_,_,_,_,_,_,_] := 0;
walk[{4,a_},_,_,_,0,_,_,_,{1,_},_,_,_,_,_] := {RND,a};
walk[{4,a_},_,_,_,0,_,_,_,{3,_},_,_,_,_] := {RND,a};
walk[{4,a_},_,_,_,0,_,_,_,_,_,_,_,{2,_}] := {RND,a};
walk[{4,a_},_,_,_,0,_,_,_,_,_,_,_,_] := 0;
walk[{x_,a_},_,_,_,_,_,_,_,_,_,_,_,_] := {RND,a};
walk[0,{3,_},{4,_},_,_,_,_,_,_,_,_,_,_] := 0;
walk[0,{3,_},_,{1,_},_,_,_,_,_,_,_,_,_] := 0;
walk[0,{3,_},_,_,{2,_},_,_,_,_,_,_,_,_] := 0;
walk[0,_,{4,_},{1,_},_,_,_,_,_,_,_,_,_] := 0;
walk[0,_,{4,_},_,{2,_},_,_,_,_,_,_,_,_] := 0;
walk[0,_,_,{1,_},{2,_},_,_,_,_,_,_,_,_] := 0;
walk[0,{3,a_},_,_,_,_,_,_,_,_,_,_,_] := {RND,a};
walk[0,_,{4,a_},_,_,_,_,_,_,_,_,_,_] := {RND,a};
walk[0,_,_,{1,a_},_,_,_,_,_,_,_,_,_] := {RND,a};
walk[0,_,_,_,{2,a_},_,_,_,_,_,_,_,_] := {RND,a};
walk[0,_,_,_,_,_,_,_,_,_,_,_,_] := 0;
Moore[func_, lat_] :=
  MapThread[func, Map[RotateRight[lat, #]&,
        {{0, 0}, {1, 0}, {0, -1}, {-1, 0}, {0, 1},
         {1, -1}, {-1, -1}, {-1, 1}, {1, 1}}], 2];
GN[func_, lat_] :=
  MapThread[func, Map[RotateRight[lat, #]&,
        {{0, 0}, {1, 0}, {0, -1}, {-1, 0}, {0, 1},
         {1, -1}, {-1, -1}, {-1, 1}, {1, 1}, {2, 0},
         {0, -2}, {-2, 0}, {0, 2}}], 2];
NestList[GN[walk, Moore[movestay, #]]&, society, t]]
```

C.4.6 printCounter and printMemory

```
printCounter[lat_, counter_] :=
  WriteString["stdout", counter, " "];

printMemory[lat_, counter_] :=
 (Print[];
  Print["After ", counter, " steps:"];
  Print["Date:            ", Date[]];
  Print["MemoryInUse:     ", MemoryInUse[]];
  Print["MaxMemoryUsed:   ", MaxMemoryUsed[]];
  Print[]);
```

Index

REGISTRATION CARD

Since this field is fast-moving, we expect updates and changes to occur that might necessitate sending you the most current pertinent information by paper, electronic media, or both, regarding *Simulating Society: A Mathematica® Toolkit for Modeling Socioeconomic Behavior*. Therefore, in order to not miss out on receiving your important update information, please fill out this card and return it to us promptly. Thank you.

Name: _____

Title: _____

Company: _____

Address: _____

City: _____ State: ____ Zip: _____

E-mail: _____

Areas of Interest/Technical Expertise: _____

Comments on this Publication: _____
_____
_____

❏ Please check this box to indicate that we may use your comments in our promotion and advertising for this publication.

Purchased from: _____

Date of Purchase: _____

❏ Please add me to your mailing list to receive updated information on *Simulating Society: A Mathematica® Toolkit for Modeling Socioeconomic Behavior* and other TELOS publications.

I have a(n) ❏ IBM compatible ❏ Macintosh ❏ Unix ❏ Other
Designate specific model: _____

NO POSTAGE
NECESSARY
IF MAILED
IN THE
UNITED STATES

BUSINESS REPLY MAIL
FIRST-CLASS MAIL   PERMIT NO. 5863   NEW YORK, NY

POSTAGE WILL BE PAID BY ADDRESSEE

THE
ELECTRONIC
LIBRARY
OF
SCIENCE

TELOS PROMOTION
SPRINGER-VERLAG NEW YORK, INC.
ATTN: J. Roth
175 FIFTH AVENUE
NEW YORK NY 10160-0266